Lecture Notes in Mathematics 2070

Editors:
J.-M. Morel, Cachan
B. Teissier, Paris

T0240528

For further volumes:
http://www.springer.com/series/304

Alla Detinko • Dane Flannery
Eamonn O'Brien
Editors

Probabilistic Group Theory, Combinatorics, and Computing

Lectures from the Fifth de Brún Workshop

Springer

Editors
Alla Detinko
Department of Mathematics
National University of Ireland, Galway
Ireland

Dane Flannery
Department of Mathematics
National University of Ireland, Galway
Ireland

Eamonn O'Brien
Department of Mathematics
University of Auckland
Auckland, New Zealand

ISBN 978-1-4471-4813-5   ISBN 978-1-4471-4814-2 (eBook)
DOI 10.1007/978-1-4471-4814-2
Springer London Heidelberg New York Dordrecht

Lecture Notes in Mathematics ISSN print edition: 0075-8434
                         ISSN electronic edition: 1617-9692

Library of Congress Control Number: 2012956221

Mathematics Subject Classification (2010): 05-04; 05B05; 20B40; 20D06; 20P05

Printed on acid-free paper

Springer is part of Springer Science+Business Media (www.springer.com)

Preface

This book is inspired by the workshop *Groups, Combinatorics, Computing*, held at National University of Ireland, Galway from April 11 to 16, 2011—the Fifth "de Brún Workshop" run under the auspices of Science Foundation Ireland's Mathematics Initiative Programme. A principal theme of the workshop was interactions between group theory and combinatorics with algorithmic or computational aspects. Areas encompassed by this theme are currently the focus of intense research activity.

The core part of the workshop was formed by three lecture courses. These contained a wide and unique selection of material, for the first time providing an accessible introduction to frontier research in thematic areas. It became clear that the courses should be made available to a larger audience.

The book has three chapters, one per lecture course. Each chapter is self-contained; beginning with background material including historical roots, the reader is led to the latest results and open problems. Illustrative examples, some proofs and algorithms, and extensive bibliographies are given.

The first chapter, by Martin Liebeck, is an exposition of recent developments in probabilistic and asymptotic theory of finite groups, particularly finite simple groups. The first two sections are on random generation of finite groups and maximal subgroups. The next topic is representation varieties and character-theoretic methods. Finally, diameter and growth of Cayley graphs of simple groups are considered. The chapter traces progress on fundamental conjectures which have driven the development of this subject.

The second chapter is by Alice Niemeyer, Cheryl Praeger, and Ákos Seress. This chapter again has a strong probabilistic flavour. It discusses the role of estimation in the design and analysis of randomised algorithms for computing with finite groups, and approaches to estimating proportions of important element classes. Among the latter are geometric methods, the use of generating functions, and theory of Lie type groups. The chapter also surveys numerous results concerning estimation in permutation groups and finite classical groups. An application to the construction of involution centralisers, a key part of the constructive recognition of finite simple groups, is given. Connections with theoretical computer science are made.

In the final chapter, Leonard Soicher presents results from a different area at the interface of group theory and combinatorics. This chapter emphasises practical computation. Specifically, it considers how group theory may be used in the construction, classification, and analysis of combinatorial designs. Statistical optimality results for semi-Latin squares are reviewed. An account of the new theory of "uniform" semi-Latin squares and a construction which determines a semi-Latin square from a transitive permutation group are then given. The chapter describes use of the GAP package DESIGN. Along with an introduction to the package and samples of its operation, it is shown how package functions can be used to classify block designs and semi-Latin squares. In an extended example, new statistically efficient semi-Latin squares are determined.

We envisage that this book will be a resource for lecture or reading courses or for self-instruction. Indeed, each chapter is a ready-made graduate lecture course. All three chapters could serve as the foundation of an advanced graduate programme in algebra and computing.

The Fifth de Brún Workshop also featured research talks and short presentations. More details and pdf files of selected talks are available at

http://www.maths.nuigalway.ie/~detinko/DeBrun5/

We take this opportunity to record our gratitude to Charles Leedham-Green, who acted as scientific chair of the workshop. He gave the opening address, and his many other contributions helped to ensure the event's success.

In conclusion, we hope that the reader will find these lectures as interesting and valuable as workshop participants did.

Galway, Ireland   Alla Detinko and Dane Flannery
Auckland, New Zealand   Eamonn O'Brien
January 2012

Acknowledgements

We thank the authors for agreeing to publication of their courses, and for their efforts in preparing the courses for this book. The useful and prompt feedback from our referees is also very much appreciated.

The workshop received funding from Science Foundation Ireland grant 07/MI/ 007. NUI Galway provided further support. We are especially indebted to Mary Kelly and Padraig Ó Catháin, for assistance with many tasks involved in running the workshop.

Contents

Contributors

Martin W. Liebeck Department of Mathematics, Imperial College, London, UK

Alice C. Niemeyer Centre for the Mathematics of Symmetry and Computation, School of Mathematics and Statistics, The University of Western Australia, Crawley, WA, Australia

Cheryl E. Praeger Centre for the Mathematics of Symmetry and Computation, School of Mathematics and Statistics, The University of Western Australia, Crawley, WA, Australia

King Abdulaziz University, Jeddah, Saudi Arabia

Ákos Seress Centre for the Mathematics of Symmetry and Computation, School of Mathematics and Statistics, The University of Western Australia, Crawley, WA, Australia

The Ohio State University, Columbus, OH, USA

Leonard H. Soicher School of Mathematical Sciences, Queen Mary University of London, London, UK

Chapter 1
Probabilistic and Asymptotic Aspects of Finite Simple Groups

Martin W. Liebeck

This is a survey of recent developments in the probabilistic and asymptotic theory of finite groups, with an emphasis on the finite simple groups. The first two sections are concerned with random generation, while the third section focusses on some applications of probabilistic methods in representation theory. The final section deals with asymptotic aspects of the diameter and growth of Cayley graphs.

1.1 Random Generation of Simple Groups and Maximal Subgroups

1.1.1 Alternating Groups

It is an elementary and well known fact that every alternating group A_n can be generated by two elements—for example, by $(1\,2\,3)$ and $(1\,2\cdots n)$ if n is odd, and by $(1\,2\,3)$ and $(2\cdots n)$ if n is even. As long ago as 1892, Netto conjectured that almost all pairs of elements of A_n will generate the whole group (see [78, p. 90]). That is, if for a finite group G we define $P(G)$ to be the probability that $\langle x, y \rangle = G$ for $x, y \in G$ chosen uniformly at random—so that

$$P(G) = \frac{|\{(x, y) \in G \times G : \langle x, y \rangle = G\}|}{|G|^2},$$

then Netto's conjecture was that $P(A_n) \to 1$ as $n \to \infty$.

It was not until 1969 that Netto's conjecture was proved by Dixon [16]:

M.W. Liebeck (✉)
Department of Mathematics, Imperial College, London SW7 2BZ, UK
e-mail: m.liebeck@imperial.ac.uk

A. Detinko et al. (eds.), *Probabilistic Group Theory, Combinatorics, and Computing*,
Lecture Notes in Mathematics 2070, DOI 10.1007/978-1-4471-4814-2_1,
© Springer-Verlag London 2013

Theorem 1.1. *Netto's conjecture holds—that is, $P(A_n) \to 1$ as $n \to \infty$.*

In fact Dixon proved more than this, showing that $P(A_n) > 1 - \frac{8}{(\log\log n)^2}$ for sufficiently large n.

We shall give a sketch of Dixon's proof. It relies on two classical results on permutation groups. The first goes back to Jordan (1873); see [18, p. 84] for a proof.

Lemma 1.2. *Suppose that X is a subgroup of S_n which acts primitively on the set $\{1,\ldots,n\}$ and contains a p-cycle for some prime $p \leq n - 3$. Then $X = A_n$ or S_n.*

The second result was deduced by Dixon (see Lemma 3 of [16]) from the paper [20] of Erdös and Turán—the second of their series of seven pioneering papers on the statistical theory of the symmetric group.

Lemma 1.3. *Let p_n be the probability that a permutation in S_n, chosen uniformly at random, has one of its powers equal to a p-cycle for some prime $p \leq n-3$. Then $p_n \to 1$ as $n \to \infty$.*

Sketch proof of Dixon's Theorem 1.1 Let $G = A_n$. Observe that if $x, y \in G$ do not generate G, then they both lie in a maximal subgroup M of G. Given M, the probability that this happens (for random x, y) is $\frac{|M|^2}{|G|^2} = |G : M|^{-2}$. Hence

$$1 - P(G) = \text{Prob}(\langle x, y \rangle \neq G \text{ for random } x, y) \leq \sum_{M \max G} |G : M|^{-2} \quad (1.1)$$

where the notation $M\ max\ G$ means M is a maximal subgroup of G. Let $\mathcal{M}$ be a set of representatives of the conjugacy classes of maximal subgroups of G. For a maximal subgroup M, the number of conjugates of M is $|G : N_G(M)| = |G : M|$, and hence

$$1 - P(G) \leq \sum_{M \in \mathcal{M}} |G : M|^{-1}. \quad (1.2)$$

The maximal subgroups of $G = A_n$ fall into three categories, according to their actions on the set $\{1,\ldots,n\}$:

(a) Intransitive subgroups $M = (S_k \times S_{n-k}) \cap G$ for $1 \leq k \leq [\frac{n}{2}]$.
(b) Imprimitive subgroups $M = (S_r \text{ wr } S_{n/r}) \cap G$, preserving a partition into $\frac{n}{r}$ r-subsets, for divisors r of n with $1 < r < n$.
(c) Primitive subgroups M.

Denote the contributions to the sum in (1.2) from categories (a), (b), (c) by $\Sigma_a, \Sigma_b, \Sigma_c$, respectively, so that $1 - P(G) \leq \Sigma_a + \Sigma_b + \Sigma_c$. The index in G of a maximal subgroup in (a) is $\binom{n}{k}$, and hence

$$\Sigma_a = \sum_{k=1}^{[n/2]} \binom{n}{k}^{-1}.$$

Similarly

$$\Sigma_b = \sum_{r|n, 1 < r < n} \left(\frac{n!}{(r!)^{n/r} (\frac{n}{r}!)} \right)^{-1}.$$

Elementary arguments (see Lemmas 1 and 2 in [16]) yield

$$\Sigma_a = \frac{1}{n} + O(\frac{1}{n^2}), \ \Sigma_b < n \cdot 2^{-n/4}.$$

So clearly

$$\Sigma_a + \Sigma_b \to 0 \text{ as } n \to \infty. \tag{1.3}$$

Estimating Σ_c is less straightforward, however. In fact, instead of estimating this we deal with the probability P_c that a random pair $x, y \in G$ generates a proper primitive subgroup of G. Here Lemmas 1.2 and 1.3 show that $P_c \to 0$ as $n \to \infty$. Since $1 - P(G) \le \Sigma_a + \Sigma_b + P_c$, it follows from (1.3) that $1 - P(G) \to 0$ as $n \to \infty$, proving Dixon's theorem. □

The right hand sides of the inequalities (1.1), (1.2) suggest that we define, for any finite group G, the *maximal subgroup zeta function*

$$\zeta_G(s) = \sum_{M \text{ max } G} |G : M|^{-s} = \sum_{n \ge 1} m_n(G) n^{-s}$$

for a real variable s, where $m_n(G)$ denotes the number of maximal subgroups of index n in G. By the argument for (1.1), we have

$$P(G) \ge 1 - \zeta_G(2). \tag{1.4}$$

For $G = A_n$ we expressed $\zeta_G(2) = \Sigma_a + \Sigma_b + \Sigma_c$ in the above proof. Using the classification of finite simple groups (CFSG), Babai showed in [4] that the number of conjugacy classes of maximal primitive subgroups of A_n is at most $c^{\log^4 n}$ for some absolute constant c. Each such subgroup has index at least $\frac{1}{2}[\frac{n+1}{2}]!$ by a classical result of Bochert (see [18, Theorem 3.3B]), and hence

$$\Sigma_c \le 2c^{\log^4 n} ([\frac{n+1}{2}]!)^{-1} = O(\frac{1}{n^2}).$$

This proves [4, 1.2]:

Theorem 1.4. *For $G = A_n$ we have $\zeta_G(2) = \frac{1}{n} + O(\frac{1}{n^2})$ and $P(G) = 1 - \frac{1}{n} + O(\frac{1}{n^2})$.*

A detailed asymptotic expansion of $P(A_n)$ can be found in [17], and precise bounds are given in [73].

1.1.2   Groups of Lie Type

At this point we move on to discuss the other non-abelian finite simple groups. By the classification (CFSG), these are the finite groups of Lie type, together with the 26 sporadic groups. Steinberg proved in [84] that every simple group of Lie type is 2-generated (i.e. can be generated by two elements), and this has also been verified for the sporadic groups (see [2]). Hence $P(G) > 0$ for all finite simple groups G. In the same paper [16] in which he proved Theorem 1.1, Dixon also made the following conjecture:

Dixon's Conjecture. *For finite simple groups G we have $P(G) \to 1$ as $|G| \to \infty$.*

For alternating groups this is of course the content of Theorem 1.1. The conjecture was proved for classical groups by Kantor and Lubotzky in [37], and for exceptional groups of Lie type by Liebeck and Shalev in [50]. These proofs were rather lengthy, but more recent work has led to a much shorter proof, which we shall now sketch. It is based on the following result, taken from [58].

Theorem 1.5. *Fix $s > 1$. For finite simple groups G, we have $\zeta_G(s) \to 0$ as $|G| \to \infty$.*

Note that the condition $s > 1$ in the theorem is necessary, since as in (1.2) above,

$$\zeta_G(s) = \sum_{M \in \mathcal{M}} |G : M|^{1-s} \tag{1.5}$$

where $\mathcal{M}$ is a set of representatives of the conjugacy classes of maximal subgroups of G. Before discussing the proof of the theorem, here are a couple of immediate consequences. Firstly, by (1.4) we have

Corollary 1.6. *Dixon's conjecture holds.*

Next, recalling that $\zeta_G(s) = \sum_{n \geq 1} m_n(G) n^{-s}$ where $m_n(G)$ denotes the number of maximal subgroups of index n in G, we deduce

Corollary 1.7. *Given $\epsilon > 0$, there exists N such that for any $n > N$ and any finite simple group G, we have $m_n(G) < n^{1+\epsilon}$.*

Sketch proof of Theorem 1.5

For $G = A_n$, the argument given for $\zeta_G(2)$ in the proof of Theorem 1.4 works equally well replacing 2 by any $s > 1$, to give $\zeta_G(s) = O(n^{1-s})$. So the main task is to prove Theorem 1.5 for groups of Lie type.

Classical Groups

We begin with an elementary example.

Example. Let $G = PSL_2(q)$ with q odd. It is well known (see for example [33, p. 191]) that the maximal subgroups of G are among the following: P (a parabolic subgroup of index $q + 1$), $D_{q\pm 1}$ (dihedral), $PSL_2(q_0)$ or $PGL_2(q_0)$ (where $\mathbb{F}_{q_0}$ is a subfield of $\mathbb{F}_q$), A_4, S_4 or A_5. There are at most 2 conjugacy classes for each subgroup listed, and the number of subfields of $\mathbb{F}_q$ is at most $\log_2 \log_2 q$. It follows that for $s > 1$ we have $\zeta_G(s) = (q + 1)^{1-s} + O(q^{\frac{3}{2}(1-s)} \log\log q) = O(q^{1-s})$. In particular, $\zeta_G(s) \to 0$ as $q \to \infty$.

Now we sketch the general argument of the proof of Theorem 1.5 for classical groups, which is essentially that given in [37]. Let $G = Cl_n(q)$ denote a simple classical group over $\mathbb{F}_q$ with natural module V of dimension n (so $G = PSL_n(q)$, $PSU_n(q)$, $PSp_n(q)$ or $P\Omega_n^\epsilon(q)$). According to a well known theorem of Aschbacher [1], the maximal subgroups of G fall into the following nine families:

$\mathscr{C}_1$: Stabilizers of totally singular or nonsingular subspaces of V (any subspaces if $G = PSL_n(q)$).
$\mathscr{C}_2$: Stabilizers of direct sum decompositions of V.
$\mathscr{C}_3$: Stabilizers of extension fields of $\mathbb{F}_q$ of prime degree.
$\mathscr{C}_4$: Stabilizers of tensor product decompositions $V = V_1 \otimes V_2$.
$\mathscr{C}_5$: Stabilizers of subfields of $\mathbb{F}_q$ of prime index.
$\mathscr{C}_6$: Normalizers of r-groups of symplectic type in absolutely irreducible representations (r a prime not dividing q).
$\mathscr{C}_7$: Stabilizers of tensor product decompositions $V = V_1 \otimes \cdots \otimes V_m$ with all V_i isometric.
$\mathscr{C}_8$: Classical subgroups (of type $PSp_n(q)$, $PSO_n(q)$, $PSU_n(q^{1/2})$ in $G = PSL_n(q)$, or $O_n(q)$ in $Sp_n(q)$ with q even).
$\mathscr{S}$: Almost simple subgroups with socle acting absolutely irreducibly on V and defined over no proper subfield of $\mathbb{F}_q$ (of $\mathbb{F}_{q^2}$ if G is unitary).

Define $\mathscr{C} = \bigcup_{i=1}^8 \mathscr{C}_i$, and denote by $N_\mathscr{C}$ (respectively, $N_\mathscr{S}$) the total number of G-conjugacy classes of maximal subgroups in $\mathscr{C}$ (respectively, in $\mathscr{S}$).

Lemma 1.8. *Let $G = Cl_n(q)$ as above. There are positive absolute constants c_1, c_2, c_3, c_4 such that the following hold:*

(i) $N_\mathscr{C} < c_1 n^2 + n \log\log q$;
(ii) $N_\mathscr{S} \leq f(n)$, where $f(n)$ is a function depending only on n;
(iii) also $N_\mathscr{S} \leq c_2 n^2 q^{6n} \log q$; and $|G : M| > q^{c_3 n^2}$ for all $M \in \mathscr{S}$;
(iv) $|G : M| > c_4 q^{n/2}$ for all maximal subgroups M of G.

Precise descriptions of the families $\mathscr{C}_i$ can be found in [40, Chap. 4], and the number of conjugacy classes of subgroups in each family is also given there. Adding these numbers up, we easily see that the number of G-conjugacy classes of maximal subgroups in $\bigcup_{i\neq 5} \mathscr{C}_i$ is less than $c_1 n^2$ for some absolute constant c_1, while the number in $\mathscr{C}_5$ is less than $n \log\log q$. Part (i) of the lemma follows.

Short arguments for parts (ii) and (iii) can be found in [58, p. 552] and [51, p. 89]. Finally, (iv) follows from [40, 5.2.2] (which gives the subgroups of minimal index in all simple classical groups).

Corollary 1.9. *The total number of conjugacy classes of maximal subgroups of* $Cl_n(q)$ *is at most* $f(n) + c_1 n^2 + n \log \log q$.

Remark. It is of interest to find good bounds for the function $f(n)$ in part (ii) of the lemma. This involves estimating, for each finite quasisimple group S, prime p and natural number n, the number of absolutely irreducible representations of S of dimension n over a field of characteristic p. This is a tough problem, especially when S is an alternating group or a group of Lie type in characteristic p. A recent paper [25] of Guralnick, Larsen and Tiep bounds the number of absolutely irreducible representations of any S of dimension n by the function $n^{3.8}$, and uses this to show that $f(n) < an^6$ for some absolute constant a. This has been sharpened by Häsä [29], who has shown that the total number $m(G)$ of conjugacy classes of maximal subgroups of any almost simple group G with socle a classical group $Cl_n(q)$ satisfies

$$m(G) < 2n^{5.2} + n \log_2 \log_2 q.$$

Using Lemma 1.8 we can complete the proof of Theorem 1.5 for classical groups $G = Cl_n(q)$. Let $s > 1$. The argument is divided into the cases where n is bounded and where n is unbounded. For the case where n is bounded, parts (i), (ii) and (iv) of the lemma give

$$\zeta_G(s) = \sum_{M \in \mathcal{M}} |G : M|^{1-s} < (f(n) + c_1 n^2 + n \log \log q)(c_4 q^{n/2})^{1-s}$$

which tends to 0 as $q \to \infty$. And for the case where n is unbounded, parts (i), (iii) and (iv) of the lemma give

$$\zeta_G(s) \leq \sum_{reps.\, M \in \mathcal{C}} |G : M|^{1-s} + \sum_{reps.\, M \in \mathcal{S}} |G : M|^{1-s}$$
$$\leq (c_1 n^2 + n \log \log q)(c_4 q^{n/2})^{1-s} + (c_2 n^2 q^{6n} \log q)(q^{c_3 n^2})^{1-s}$$

which tends to 0 as $n \to \infty$.

Exceptional Groups of Lie Type

Now we discuss the proof of Theorem 1.5 when $G = G(q)$ is a simple exceptional group of Lie type over $\mathbb{F}_q$—that is, a group in one of the families $E_8(q)$, $E_7(q)$, $E_6(q)$, $^2E_6(q)$, $F_4(q)$, $^2F_4(q)$, $G_2(q)$, $^3D_4(q)$, $^2G_2(q)$, $^2B_2(q)$.

Let $\bar{G}$ be the simple algebraic group over $K = \bar{\mathbb{F}}_q$, the algebraic closure of $\mathbb{F}_q$, corresponding to $G(q)$; so if $G = E_8(q)$ then $\bar{G} = E_8(K)$, if $G = {}^2E_6(q)$ then $\bar{G} = E_6(K)$, and so on. There is a Frobenius endomorphism σ of $\bar{G}$ such that $G = \bar{G}'_\sigma$, where $\bar{G}_\sigma$ denotes the fixed point group $\{g \in \bar{G} : g^\sigma = g\}$ (see [85]). When G is not a twisted group, σ is just a field morphism which acts on root groups $U_\alpha = \{U_\alpha(t) : t \in K\}$ as $U_\alpha(t) \to U_\alpha(t^q)$; when G is twisted, σ also involves a graph morphism of $\bar{G}$.

Beginning with the work of Dynkin [19], a great deal of effort has gone into determining the maximal closed subgroups of positive dimension in the algebraic group $\bar{G}$, culminating in [49], where this task was completed. The conclusion (see [49, Corollary 2]) is that there are only finitely many conjugacy classes of maximal closed subgroups of positive dimension in $\bar{G}$:

(a) Maximal parabolic subgroups.
(b) Normalizers of reductive subgroups of maximal rank—these are subgroups containing a maximal torus of $\bar{G}$, and have root system a subsystem of the root system of $\bar{G}$.
(c) A few further classes of (normalizers of) semisimple subgroups.

For example, when $\bar{G} = E_6(K)$ the subgroups under (a) are the parabolics P_i for $1 \le i \le 6$; those under (b) are the normalizers of the subsystem subgroups $A_1 A_5$, A_2^3, $D_4 T_2$ and T_6 (a maximal torus); and those under (c) are the normalizers of subgroups of types F_4, C_4, A_2, G_2, and $A_2 G_2$.

In parallel with this, there are results which relate the subgroup structure of the finite groups $G(q)$ with that of $\bar{G}$. The following is taken from [47, Corollary 8]:

Theorem 1.10. *There is an absolute constant c such that if M is a maximal subgroup of the exceptional group $G(q) = \bar{G}_\sigma'$, then one of the following holds:*

(i) $|M| < c$;
(ii) *M is a subfield subgroup $G(q_0)$, or a twisted subgroup (such as ${}^2E_6(q^{1/2}) < E_6(q)$);*
(iii) *$M = \bar{M}_\sigma$ for some σ-stable maximal closed subgroup $\bar{M}$ of $\bar{G}$ of positive dimension.*

The maximal subgroups M under (ii), (iii) fall into at most $d \log \log q$ conjugacy classes of subgroups in $G(q)$; all satisfy $|G(q) : M| > d'q^r$ (here r is the rank of $\bar{G}$ and d, d' are positive absolute constants).

This is nice, but it gives no information about the number of conjugacy classes of bounded maximal subgroups under (i). The possibilities for these subgroups were determined up to isomorphism in [48], but nothing much was proved about their conjugacy until the work of Ben Martin [74], which was a major ingredient in the proof of the following result, taken from [58, Theorem 1.2]:

Theorem 1.11. *Let N, R be positive integers, and let G be a finite almost simple group with socle a group of Lie type of rank at most R. Then the number of conjugacy classes of maximal subgroups of G of order at most N is bounded by a function $f(N, R)$ of N, R only.*

Applying this to the finite exceptional groups of Lie type, and combining with Theorem 1.10, we have:

Corollary 1.12. *There is an absolute constant e such that the number of conjugacy classes of maximal subgroups of any finite exceptional group $G(q)$ is bounded above by $e \log \log q$.*

This leads immediately to the proof of Theorem 1.5 for exceptional groups: for $s > 1$,

$$\zeta_G(s) = \sum_{reps.\, M} |G : M|^{1-s} < (e \log \log q)(d'q^r)^{1-s}$$

which tends to 0 as $q \to \infty$.   □

We conclude this section with a brief discussion of Theorem 1.11. The proof of this uses geometric invariant theory, via the theory of *strongly reductive* subgroups of the algebraic group $\bar{G}$, a notion due to Richardson. These are closed subgroups H which are not contained in any proper parabolic subgroup of $C_{\bar{G}}(T)$, where T is a maximal torus of $C_{\bar{G}}(H)$; in particular, subgroups lying in no proper parabolic of $\bar{G}$ are strongly reductive. Martin's main result in [74] is that the number of conjugacy classes of strongly reductive subgroups of $\bar{G}$ of order at most N is bounded by a function $g(N, R)$, where R is the rank of $\bar{G}$. To adapt this to the analysis of subgroups of the finite group $G(q) = \bar{G}_\sigma$, the following was proved in [58, 2.2]:

Lemma 1.13. *If F is a finite subgroup of $\bar{G}$ which is invariant under σ, then either F is strongly reductive, or F is contained in a σ-invariant proper parabolic subgroup of $\bar{G}$.*

The proof of this involves geometric invariant theory. The lemma implies that maximal non-parabolic subgroups of $\bar{G}_\sigma$ are strongly reductive in $\bar{G}$, at which point Martin's result can be used to deduce Theorem 1.11.

We mention finally that Corollaries 1.9 and 1.12 can be used along with arguments in [52] to prove the following, which is [58, 5.2]:

Theorem 1.14. *The symmetric group has $n^{o(1)}$ conjugacy classes of primitive maximal subgroups, and $\frac{1}{2}n + n^{o(1)}$ classes of maximal subgroups in total.*

1.1.3 Other Results on Random Generation

We have discussed at length Dixon's conjecture that for a finite simple group G, $P(G) = \mathrm{Prob}(\langle x, y \rangle = G) \to 1$ as $|G| \to \infty$. Since the conjecture was proved, many variants have been established where one insists on various properties of the generators x, y. We now briefly discuss some of these.

The first concerns the notion of $(2, 3)$-generation. A group G is said to be $(2, 3)$-*generated* if it can be generated by two elements x, y such that $x^2 = y^3 = 1$. It is well known that such groups are precisely the images of the modular group $PSL_2(\mathbb{Z})$ (since $PSL_2(\mathbb{Z})$ is isomorphic to the free product of the cyclic groups C_2 and C_3). A question which goes back a long way is:

Problem. *Which finite simple groups are $(2, 3)$-generated?*

For simple alternating groups this was answered in 1901 by G.A. Miller [77], who showed that A_n is $(2, 3)$-generated if and only if $n \neq 6, 7, 8$. There is quite a large literature on the problem for classical groups—for example, Tamburini et al. [88]

showed that many classical groups of large dimension are $(2, 3)$-generated. The approach in these papers and many others is to produce explicit generators of the required orders. Is there a probabilistic approach?

In [51], for any finite group G, Liebeck and Shalev defined $P_{2,3}(G)$ to be the probability that two randomly chosen elements x, y of orders 2, 3 generate G. That is, writing $I_r(G)$ for the set of elements of order r in G, and $i_r(G) = |I_r(G)|$,

$$P_{2,3}(G) = \frac{|\{(x, y) \in I_2(G) \times I_3(G) : \langle x, y \rangle = G\}|}{i_2(G)i_3(G)}.$$

For finite simple groups G, we can attempt to estimate $P_{2,3}(G)$ using a similar approach to the proof of Dixon's conjecture. Given a maximal subgroup M of G, the probability that a random pair x, y of elements of orders 2, 3 lies in M is $\frac{i_2(M)i_3(M)}{i_2(G)i_3(G)}$, and hence

$$1 - P_{2,3}(G) \leq \sum_{M \, \max G} \frac{i_2(M)i_3(M)}{i_2(G)i_3(G)}.$$

We were able to prove that for alternating groups, and also for classical groups with some low-dimensional exceptions,

$$\frac{i_2(M)}{i_2(G)} < c|G : M|^{-2/5}, \quad \frac{i_3(M)}{i_3(G)} < c|G : M|^{-5/8}$$

for all maximal subgroups M, where c is a constant. Hence, excluding the low-dimensional exceptions,

$$1 - P_{2,3}(G) \leq \sum_{M \, \max G} c^2|G : M|^{-2/5-5/8} = c^2 \zeta_G(\tfrac{41}{40}),$$

which tends to 0 as $|G| \to \infty$ by Theorem 1.5. In fact the low-dimensional exceptions led to some interesting and unexpected counterexamples to $(2, 3)$-generation, namely four-dimensional symplectic groups in characteristics 2 and 3. The final result is [51, 1.4]:

Theorem 1.15. *For G an alternating group, or a finite simple classical group not isomorphic to $PSp_4(q)$, we have $P_{2,3}(G) \to 1$ as $|G| \to \infty$.*

For $G = PSp_4(2^a)$ or $PSp_4(3^a)$ we have $P_{2,3}(G) = 0$; while for $G = PSp_4(p^a)$ with $p > 3$ prime, we have $P_{2,3}(G) \to \frac{1}{2}$ as $p^a \to \infty$.

As a consequence, all but finitely many classical groups ($\neq PSp_4(2^a)$, $PSp_4(3^a)$) are $(2, 3)$-generated. The exceptional families $PSp_4(2^a)$, $PSp_4(3^a)$ were a surprise at the time, but are rather easily seen not to be $(2, 3)$-generated. For example, consider $G = PSp_4(q)$ with $q = 3^a$, and regard G as the isomorphic orthogonal group $\Omega_5(q) = \Omega(V)$. If $x, y \in G$ are elements of orders 2, 3 respectively, then the -1-eigenspace of x on V has dimension at least 3, while the 1-eigenspace of y has dimension 3. Hence these eigenspaces intersect nontrivially, and it follows that $\langle x, y \rangle \neq G$.

There are many further results of this flavor in the literature. Here is a selection. For a positive integer k, define $P_{k,*}(G)$ to be the probability that G is generated by a random element of order k and a random further element; so

$$P_{k,*}(G) = \frac{|\{(x, y) \in I_k(G) \times G : \langle x, y \rangle = G\}|}{i_k(G)|G|}.$$

Let $P_C(G)$ be the probability that G is generated by two randomly chosen conjugates—that is, $P_C(G) = \mathrm{Prob}(\langle x, x^y \rangle = G$ for random $x, y \in G)$.

Theorem 1.16. *For all finite simple groups G,*

  (i) $P_{2,*}(G) \to 1$ *as* $|G| \to \infty$
 (ii) $P_{3,*}(G) \to 1$ *as* $|G| \to \infty$
(iii) $P_C(G) \to 1$ *as* $|G| \to \infty$.

Parts (i) and (ii) are taken from [53], while (iii) is the main result of [27].

We conclude this section by mentioning a couple of results of a slightly different and rather useful style, in that they provide explicit constants. The first is taken from [24]:

Theorem 1.17. *In any finite simple group G there is a conjugacy class C such that for any $1 \neq x \in G$,*

$$\mathrm{Prob}(\langle x, c \rangle = G \text{ for random } c \in C) > \frac{1}{10}.$$

In other words, for every non-identity $x \in G$, we have $\langle x, c \rangle = G$ for at least a tenth of the elements $c \in C$. As a consequence, for every $1 \neq x \in G$ there exists y such that $\langle x, y \rangle = G$ (a property known as the $\frac{3}{2}$-*generation* of G).

The final result is taken from [76]:

Theorem 1.18. *We have $P(G) \geq \frac{53}{90}$ for all finite simple groups G, with equality if and only if $G = A_6$.*

Further results on random generation of simple groups (such as "random Fuchsian generation") can be found in Sect. 1.3.

1.1.4 Generation of Maximal Subgroups

Having discussed the generation of finite simple groups, we move on to the generation of their maximal subgroups. We know that $d(G) = 2$ for every (non-abelian) simple group, where $d(G)$ denotes the minimal number of generators of G. What about $d(M)$ for maximal subgroups M of G?

We have sketched some results about maximal subgroups of simple groups in the previous sections. Much more complete information can be found in [40] for classical groups and in [46] for exceptional groups of Lie type; in particular, the only unknown maximal subgroups M are almost simple, and for these we have $d(M) \leq 3$ by [15]. And for alternating groups, the O'Nan–Scott theorem (see for example the Appendix of [3]) shows that the primitive maximal subgroups fall into several classes—affine type, product type, diagonal type and almost simple type— so again the unknown ones are almost simple.

Despite being "known", the non-almost simple maximal subgroups can have quite intricate structures, which makes the proof of the following recent result, taken from [11, Theorem 1], rather delicate.

Theorem 1.19. *If G is a finite simple group, and M is a maximal subgroup of G, then $d(M) \leq 4$.*

Equality can hold here. For example, let $G = P\Omega_{a^2}^+(q)$ with $a \equiv 2 \bmod 4$ and $q \equiv 1 \bmod 4$. Then G has a maximal subgroup M of type $O_a^+ \otimes O_a^+$, in the tensor product family $\mathscr{C}_7$ (see [40, 4.7.6]), and the precise structure of M is $(P\Omega_a^+(q) \times P\Omega_a^+(q)).2^4$, so clearly $d(M) \geq 4$.

The theorem has quite a neat consequence concerning primitive permutation groups (see [11, Theorem 7]): if G is a primitive group with point-stabilizer M, then $d(M) \leq d(G) + 4$.

Having found the minimal number of generators, one might ask questions about the random generation properties of maximal subgroups. Some of these will be addressed in the next section (see Corollary 1.26).

1.2 Random Generation of Arbitrary Finite Groups

For a finite group G, recall that $d(G)$ is the minimal number of generators of G. For $k \geq d(G)$ let $d_k(G)$ be the number of generating k-tuples of elements of G. That is,

$$d_k(G) = |\{(x_1, \ldots, x_k) \in G^k : \langle x_1, \ldots, x_k \rangle = G\}|.$$

Set

$$P_k(G) = \frac{d_k(G)}{|G|^k},$$

so that $P_k(G)$ is the probability that k randomly chosen elements generate G. Following Pak [79], define

$$v(G) = \min\{k \in \mathbb{N} : P_k(G) \geq \frac{1}{e}\}$$

(where e is as usual the base of natural logarithms). The choice of the constant $1/e$ here is for convenience, and is not significant; note that for any $r \geq 1$, we have $P_{rv(G)}(G) \geq 1 - (1 - \frac{1}{e})^r$.

A basic goal is to understand the relationship between $v(G)$ and $d(G)$ for finite groups G. We begin with a couple of examples.

Examples. 1. Let G be a p-group for some prime p, and let $d = d(G)$. For any k, we have $P_k(G) = P_k(G/\Phi(G)) = P_k(C_p^d)$. Taking $k = d$,

$$P_d(G) = P_d(C_p^d) = \frac{p^d - 1}{p^d} \cdot \frac{p^d - p}{p^d} \cdots \frac{p^d - p^{d-1}}{p^d} = \prod_1^d (1 - \frac{1}{p^i}),$$

and it is easy to see that this product is at least $1 - \frac{1}{p} - \frac{1}{p^2}$, which is greater than $\frac{1}{e}$ when $p > 2$, and is $\frac{1}{4}$ when $p = 2$. A slight refinement gives $P_{d+1}(G) > \frac{1}{e}$ in all cases, and hence $v(G) \le d(G) + 1$ for p-groups.
2. For simple groups G, Dixon's conjecture says that $P_2(G) \to 1$ as $|G| \to \infty$, which implies that $v(G) = 2$ for sufficiently large G. In fact, Theorem 1.18 gives $v(G) = 2$ for all finite simple groups G.

It is not too hard to generalize Example 1 to all nilpotent groups. This is done in [79], where the much less easy case of soluble groups is also considered:

Theorem 1.20. (i) *For G nilpotent, $v(G) \le d(G) + 1$.*
(ii) *For G soluble, $v(G) \le 3.25\, d(G) + 10^7$.*

Examples due to Mann [69] show that it is not possible to improve the constant 3.25 in part (ii) by much.

One might be tempted to think that for any finite group G, $v(G)$ and $d(G)$ are closely related—perhaps $v(G) < c \cdot d(G)$ for some absolute constant c? This is in fact far from being true, as is shown by the following result, taken from [37].

Lemma 1.21. *For any real number R, there exists a finite group G such that $d(G) = 2$ and $v(G) > R$.*

Proof. To prove the lemma, Kantor and Lubotzky construct such a group G of the form T^N, where T is a non-abelian simple group. A result of Philip Hall [28] shows that the maximal value of N such that T^N is 2-generated is $d_2(T)/|\mathrm{Aut}(T)|$. For example, when $T = A_5$, Hall calculated that $d_2(A_5)/|S_5| = 19$, so that A_5^{19} is 2-generated whereas A_5^{20} is not.

Now let $T = A_n$. By Dixon's Theorem 1.1, for large n, $d_2(T)$ is at least $\frac{2}{3}|T|^2$, and so $d_2(T)/|\mathrm{Aut}(T)|$ is at least $\frac{2}{3}|A_n|^2/|S_n| = n!/6$. Set $N = n!/8$, and define $G = T^N$. Then $d(G) = 2$.

Now consider $P_k(G)$. The probability that a random k-tuple of elements of A_n generates A_n is at most $1 - \frac{1}{n^k}$ (since $\frac{1}{n^k}$ is the probability that all of the k permutations fix 1). If a random k-tuple in $G = A_n^N$ generates G, then each of the N k-tuples in a given coordinate position must generate A_n, and the probability that this happens is at most $(1 - \frac{1}{n^k})^N = (1 - \frac{1}{n^k})^{n!/8}$. For this to be at least $\frac{1}{e}$, k must be of the order of n. Hence $v(G)$ can be arbitrarily large, while $d(G) = 2$. $\qquad\square$

So it seems that it will be tricky to find a general relationship between $v(G)$ and $d(G)$. In [79], Pak proves that $v(G) \leq \lceil \log_2 |G| \rceil + 1$ for all finite groups G, and conjectures that there is a constant C such that $v(G) < C \cdot d(G) \cdot \log \log |G|$. This was proved in a strong form by Lubotzky [66] (see also [15, Theorem 20]):

Theorem 1.22. *For all finite groups G,*

$$v(G) \leq d(G) + 2 \log_2 \log_2 |G| + 4.02.$$

We now discuss Lubotzky's proof in some detail. Recall that $m_n(G)$ denotes the number of maximal subgroups of index n in the finite group G. Define

$$\mu(G) = \max_{n \geq 2} \frac{\log m_n(G)}{\log n}.$$

So $m_n(G) \leq n^{\mu(G)}$ for all n, and we can think of $\mu(G)$ as the "polynomial degree" of the rate of growth of $m_n(G)$.

Lemma 1.23. *For any finite group G, $v(G) \leq \lceil \mu(G) + 2.02 \rceil$.*

Proof. For any positive integer k, if $x_1, \ldots, x_k$ denote randomly chosen elements of G, we have

$$
\begin{aligned}
1 - P_k(G) = \operatorname{Prob}(\langle x_1, \ldots, x_k \rangle \neq G) \\
\leq \sum_{M \, max \, G} \operatorname{Prob}(x_1, \ldots, x_k \in M) \\
= \sum_{M \, max \, G} \frac{|M|^k}{|G|^k} \\
= \sum_{n \geq 2} m_n(G) n^{-k} \\
\leq \sum_{n \geq 2} n^{\mu(G) - k}.
\end{aligned}
$$

Hence if $k \geq \mu(G) + 2.02$, then $1 - P_k(G) \leq \sum_{n \geq 2} n^{-2.02}$, which is less than $1 - \frac{1}{e}$. Therefore $P_k(G) \geq \frac{1}{e}$ for such k, giving the result. □

By the lemma, to prove Theorem 1.22, it is sufficient to bound $m_n(G)$ for arbitrary finite groups G. Each maximal subgroup of index n in G gives a homomorphism $\pi : G \to S_n$ with image $\pi(G)$ a primitive subgroup of S_n and kernel $\operatorname{Ker}(\pi) = \operatorname{core}_G(M) = \bigcap_{g \in G} M^g$.

We call a maximal subgroup M *core-free* if $\operatorname{core}_G(M) = 1$. According to a result of Pyber, which appeared later in improved form as [59, Theorem 1.4], the number of core-free maximal subgroups of index n in G is at most n^2 (the improved form is $cn^{3/2}$). The proof relies on the detailed description of core-free maximal subgroups given by Aschbacher and Scott in [3].

Now consider the non-core-free maximal subgroups. Each corresponds to a homomorphism $\pi : G \to S_n$ with $\pi(G)$ primitive and $\operatorname{Ker}(\pi) \neq 1$. To compute $m_n(G)$, we need to count the number of possibilities for $\operatorname{Ker}(\pi)$ (and then multiply by n^2, by Pyber's result). To do this, we consider a chief series $1 = N_0 \leq N_1 \leq \cdots \leq N_r = G$ (so $N_i \triangleleft G$ and each N_i/N_{i-1} is minimal normal in G/N_{i-1}).

A finite group can have many chief series, but the number r, and the collection of chief factors N_i/N_{i-1}, are uniquely determined by G; as is the collection of normal subgroups $C_1, \ldots, C_r$, where $C_i = C_G(N_i/N_{i-1})$, the kernel of the action of G on N_i/N_{i-1}.

By the O'Nan–Scott theorem (see [18, Theorems 4.3B, 4.7A]), there are three possibilities for the structure of the primitive permutation group $\pi(G)$:

1. $\pi(G)$ has a unique minimal normal subgroup K, and $K \cong T^k$ for some non-abelian simple group T.
2. $\pi(G)$ has exactly two minimal normal subgroups K_1, K_2, and $K_1 \cong K_2 \cong T^k$ for some non-abelian simple group T.
3. $\pi(G)$ is an *affine* group: it has a unique minimal normal subgroup $K \cong C_p^k$ for some prime p.

In case (1), Lubotzky showed that $\mathrm{Ker}(\pi)$ must be one of the r subgroups C_i; and in case (2), $\mathrm{Ker}(\pi)$ must be $C_i \cap C_j$ for some i, j (see [66, 2.3]). Thus these cases contribute at most $\frac{1}{2}r(r+1)n^2$ to $m_n(G)$. Further argument [66, 2.5] shows that case (3) contributes at most $rn^{d(G)+2}$, and hence

Lemma 1.24. *For a finite group G which has r chief factors, $m_n(G) \le r^2 n^{d(G)+2}$.*

Since $r < \log|G|$, the lemma gives $\mu(G) \le d(G) + 2 + 2\log r \le d(G) + 2 + 2\log\log|G|$, so Theorem 1.22 follows using Lemma 1.23.

Building on Lubotzky's ideas, and adding a lot more of their own, Jaikin-Zapirain and Pyber proved the following remarkable result in [34], giving upper and lower bounds for $\nu(G)$ which are tight up to a multiplicative constant.

To state the result we need a few definitions. For a non-abelian characteristically simple group A (i.e. $A \cong T^k$ with T simple), denote by $\mathrm{rk}_A(G)$ the maximal number r such that G has a normal section which is the product of r chief factors isomorphic to A. Let $l(A)$ be the minimal degree of a faithful transitive permutation representation of A. Finally, define

$$\rho(G) = \max_A \frac{\log \mathrm{rk}_A(G)}{\log l(A)}.$$

For example, if $G = (A_n)^t$ $(n \ge 5)$, then all chief factors are isomorphic to A_n, and $\mathrm{rk}_{A_n}(G) = t$, $l(A_n) = n$, so $\rho(G) = \log t/\log n$. On the other hand, if $G = A_n \mathrm{\ wr\ } A_t$ $(t \ge 5)$, then a chief series is $1 \le (A_n)^t \le G$, and $\mathrm{rk}_A(G) = 1$ for both chief factors of G, so $\rho(G) = 0$.

Theorem 1.25. *There exist absolute constants $\alpha, \beta > 0$ such that for all finite groups G,*
$$\alpha(d(G) + \rho(G)) < \nu(G) < \beta d(G) + \rho(G).$$

If we return to the example $G = (A_n)^{n!/8}$ in the proof of Lemma 1.21, the theorem says that $\nu(G)$ is of the order of $\log(n!)/\log n$, hence of the order of n (while $d(G) = 2$). On the other hand, for the wreath product $G = A_n \mathrm{\ wr\ } A_t$, the theorem tells us that $\nu(G)$ is bounded.

Applications

We conclude this section by discussing a few applications of Theorem 1.25. The first, taken from [11], is to the random generation of maximal subgroups of finite simple groups. Recall from Theorem 1.19 that such a maximal subgroup M satisfies $d(M) \leq 4$. It is shown also in [11, 8.2] that M has at most three non-abelian chief factors. Hence $\nu(M)$ is bounded by Theorem 1.25, and so we have

Corollary 1.26. *Given $\epsilon > 0$, there exists $k = k(\epsilon)$ such that $P_k(M) > 1 - \epsilon$ for any maximal subgroup M of any finite simple group.*

One might be tempted to think, in the spirit of Dixon's conjecture, that there is a constant k such that $P_k(M) \to 1$ as $|M| \to \infty$. But this is not the case, as is shown by the maximal subgroups S_{n-2} of A_n, for which $P_k(S_{n-2}) \leq 1 - \frac{1}{2^k}$ for any k.

The second application is to linear groups [34, 9.7]. Let G be a finite subgroup of $GL_n(K)$ for some field K. A result of Fisher [22] implies that the number of non-abelian chief factors of G is less than n. Hence Theorem 1.25 gives

Corollary 1.27. *There is an absolute constant c such that if G is any finite linear group in dimension n over some field K, then $\nu(G) < c \cdot d(G) + \log n$.*

It is striking that the number of random generators does not depend on the field K.

The third application relates $\nu(G)$ to the sizes of so-called *minimal* generating sets of G—that is, generating sets S such that no proper subset of S generates G. Define $\tilde{d}(G)$ to be the maximal size of a minimal generating set of G. For example, if $G = S_n$ then $\{(1\,2), (2\,3), \ldots, (n-1\,n)\}$ is a minimal generating set of size $n-1$, and a result of Whiston [89] shows that $\tilde{d}(S_n) = n - 1$.

It is proved in [34, 9.9] that $\mathrm{rk}_A(G) \leq \tilde{d}(G)$ for any non-abelian characteristically simple group A, and hence Theorem 1.25 gives

Corollary 1.28. *There is an absolute constant c such that for any finite group G, $\nu(G) < c \cdot d(G) + \log \tilde{d}(G)$.*

This result has some significance in the analysis of the Product Replacement Algorithm for choosing random elements in a finite group, since the quantities $\nu(G)$ and $\tilde{d}(G)$ play a role in this analysis. We refer the reader to [80] for details.

1.3 Representation Varieties and Character-Theoretic Methods

If Γ is a finitely generated group, K a field and n a natural number, we call $\mathrm{Hom}(\Gamma, GL_n(K))$, the set of representations $\rho : \Gamma \to GL_n(K)$, the *representation variety* of Γ in dimension n over K. In this section we shall show how probabilistic and character-theoretic methods can be brought to bear on the study of such varieties over algebraically closed fields and also over finite fields, for a particular class of finitely generated groups Γ. We shall also consider the representation spaces $\mathrm{Hom}(\Gamma, G(K))$ and $\mathrm{Hom}(\Gamma, S_n)$, where $G(K)$ is a simple algebraic group over K, which are of interest in various contexts.

1.3.1 Fuchsian Groups

The class of finitely generated groups Γ we shall consider are the *Fuchsian* groups, i.e. finitely generated discrete groups of isometries of the hyperbolic plane. By classical work of Fricke and Klein, the orientation-preserving Fuchsian groups Γ have presentations of the following form:

$$\text{generators:} \quad a_1, b_1, \ldots, a_g, b_g \ \ \text{(hyperbolic)}$$
$$x_1, \ldots, x_d \ \ \text{(elliptic)}$$
$$y_1, \ldots, y_s \ \ \text{(parabolic or hyperbolic boundary)}$$

$$\text{relations:} \quad x_1^{m_1} = \cdots = x_d^{m_d} = 1,$$
$$x_1 \cdots x_d \, y_1 \cdots y_s [a_1, b_1] \cdots [a_g, b_g] = 1$$

where $g, d, s \geq 0$ and $m_i \geq 2$ for all i, and the *measure* $\mu(\Gamma) > 0$, where

$$\mu(\Gamma) = 2g - 2 + s + \sum_{1}^{d} \left(1 - \frac{1}{m_i}\right).$$

The number g is called the *genus* of Γ.

There are also results along the lines discussed below for non-orientation preserving Fuchsian groups, but for brevity's sake we shall not mention these.

Examples. 1. *Surface groups.* These are the groups with $s = d = 0$ and $g \geq 2$: let

$$\Gamma_g = \langle a_1, b_1, \ldots, a_g, b_g : \prod_i [a_i, b_i] = 1 \rangle$$

so that $\Gamma_g = \pi_1(S)$, the fundamental group of a surface S of genus g.

2. *Triangle groups.* These have $g = s = 0$, $d = 3$. For positive integers a, b, c define

$$T = T_{a,b,c} = \langle x, y, z : x^a = y^b = z^c = xyz = 1 \rangle,$$

where $\mu(T) = 1 - \frac{1}{a} - \frac{1}{b} - \frac{1}{c} > 0$, and call this the (a, b, c)-*triangle group*. The minimal value of μ for a triangle group is $\frac{1}{42}$, which occurs for the *Hurwitz* triangle group $T_{2,3,7}$.

3. *Free products.* When $s > 0$, Γ is a free product of cyclic groups, such as the free group F_r of rank r, or a free product $C_a * C_b = \langle x, y : x^a = y^b = 1 \rangle$ (which has $\mu = 1 - \frac{1}{a} - \frac{1}{b}$).

The Fuchsian groups for which $s = 0$ are said to be *co-compact*.

Let q be a prime power and $K = \bar{\mathbb{F}}_q$, the algebraic closure of $\mathbb{F}_q$. In the sections below, we shall discuss the representation spaces $\mathrm{Hom}(\Gamma, G)$, where Γ is a Fuchsian group and G is $GL_n(K)$, $G(K)$ (a simple algebraic group over K), $G(q)$ (a group of Lie type over $\mathbb{F}_q$), or S_n. These representation spaces have many connections with other areas, some of which we shall now point out.

The first connection is with the area of random generation. To say that a finite group G is generated by two elements is to say that there exists an epimorphism in the space $\mathrm{Hom}(F_2, G)$, where F_2 is the free group of rank 2. The probability $P(G)$ (defined in Sect. 1.1.1) that G is generated by two randomly chosen elements is then

$$P(G) = \frac{|\{\phi \in \mathrm{Hom}(F_2, G) : \phi \text{ epi}\}|}{|\mathrm{Hom}(F_2, G)|} = \mathrm{Prob}(\text{random } \phi \in \mathrm{Hom}(F_2, G) \text{ is epi}).$$

For any Fuchsian group Γ, we can similarly define

$$P_\Gamma(G) = \mathrm{Prob}(\text{random } \phi \in \mathrm{Hom}(\Gamma, G) \text{ is epi}).$$

Dixon's conjecture (Corollary 1.6) asserts that $P(G) \to 1$ as $|G| \to \infty$ for finite simple groups G, and one could hope to prove similar results for the probabilities $P_\Gamma(G)$. For example, a question posed many years ago asks which finite simple groups are $(2, 3, 7)$-generated—that is, generated by three elements of orders 2, 3 and 7 with product equal to 1. There are many results on this question in which the approach is to construct explicit generators, but one might hope to shed further light by studying the probabilities $P_\Gamma(G)$, where G is simple and Γ is the triangle group $T_{2,3,7}$. Results on this and other probabilities $P_\Gamma(G)$ will be discussed below.

Other connections concern the representation space $\mathrm{Hom}(\Gamma, S_n)$. The *subgroup growth* of Γ is measured by the growth of the function $a_n(\Gamma)$, the number of subgroups of index n in Γ, and it is an elementary fact that

$$a_n(\Gamma) = |\mathrm{Hom}_{\mathrm{trans}}(\Gamma, S_n)|/(n - 1)!,$$

where $\mathrm{Hom}_{\mathrm{trans}}(\Gamma, S_n)$ is the set of homomorphisms $\Gamma \to S_n$ which have image which is transitive on $\{1, \ldots, n\}$. Another connection is Hurwitz's theory that $\mathrm{Hom}(\Gamma, S_n)$ counts branched coverings of Riemann surfaces, where g is the genus of the surface and the images of $x_1, \ldots, x_d$ are the monodromy permutations around the branch points (see [55, Sect. 8] for details).

1.3.2 Character Theory

The connection between the sizes of the spaces $\mathrm{Hom}(\Gamma, G)$ and character theory is given by the following lemma, which goes back to Frobenius and Hurwitz. Let Γ be a co-compact Fuchsian group with generators a_i, b_i, x_i as above, and let G

be a finite group. For conjugacy classes C_i in G having representatives g_i of order dividing m_i ($1 \le i \le d$), define $\mathbf{C} = (C_1, \ldots, C_d)$ and

$$\mathrm{Hom}_{\mathbf{C}}(\Gamma, G) = \{\phi \in \mathrm{Hom}(\Gamma, G) : \phi(x_i) \in C_i \text{ for } 1 \le i \le d\}.$$

Denote by $\mathrm{Irr}(G)$ the set of irreducible (complex) characters of G.

Lemma 1.29. *With the above notation,*

$$|\mathrm{Hom}_{\mathbf{C}}(\Gamma, G)| = |G|^{2g-1}|C_1| \cdots |C_d| \sum_{\chi \in Irr(G)} \frac{\chi(g_1) \cdots \chi(g_d)}{\chi(1)^{2g-2+d}}.$$

For a proof, see [55, 3.2]. For example, applying this with $g = 0$ and $d = 3$, we obtain the well known formula of Frobenius that if C_1, C_2, C_3 are three classes in G, then the number of solutions to the equation $x_1 x_2 x_3 = 1$ for $x_i \in C_i$ is

$$\frac{|C_1||C_2||C_3|}{|G|} \sum_{\chi \in Irr(G)} \frac{\chi(g_1)\chi(g_2)\chi(g_3)}{\chi(1)}. \tag{1.6}$$

When G is a finite simple group, a major tool in the analysis of $\mathrm{Hom}(\Gamma, G)$ is provided by the *representation zeta function* of G, defined for a real variable s by

$$\zeta^G(s) = \sum_{\chi \in Irr(G)} \chi(1)^{-s}.$$

The next result is taken from [55, 1.1] for alternating groups, and from [56, 1.1, 1.2] for groups of Lie type.

Theorem 1.30. (i) *If $G = A_n$ and $s > 0$, then $\zeta^G(s) \to 1$ as $n \to \infty$.*
(ii) *If $G = G(q)$ is of fixed Lie type with Coxeter number h, and $s > \frac{2}{h}$, then $\zeta^G(s) \to 1$ as $q \to \infty$.*
(iii) *For any $s > 0$, there exists $r = r(s)$ such that for $G = G(q)$ of rank at least r, we have $\zeta^G(s) \to 1$ as $|G| \to \infty$.*

In part (ii), the *Coxeter number* h of $G(q)$ is defined to be the number of roots in the root system of $G(q)$ divided by the rank; for example, if $G = PSL_n(q)$, $PSp_{2n}(q)$ or $E_8(q)$ then h is n, $2n$ or 30, respectively. The bound $\frac{2}{h}$ is tight, in that $\zeta^{G(q)}(2/h)$ is bounded away from 1. Moreover, if we write $r_n(G)$ for the number of irreducible characters of degree n, then $\zeta^G(s) = \sum_{n \ge 1} r_n(G)n^{-s}$, so for example part (ii) implies that for $G(q)$ of fixed Lie type, $r_n(G(q)) < cn^{2/h}$ for all n, where c is an absolute constant.

Theorem 1.30 also holds for the corresponding *quasisimple* groups—that is, perfect groups G such that $G/Z(G)$ is a finite simple group.

1.3.3 Symmetric and Alternating Groups

Here we discuss some of the ramifications of the representation space $\text{Hom}(\Gamma, S_n)$. Let us begin the story with a well known result of Conder [13], inspired by the ideas of his supervisor Graham Higman: for $n \geq 168$, A_n is $(2, 3, 7)$-generated. (There is a reason for the number 168 here: A_{167} is not $(2, 3, 7)$-generated.) This was part of a more general conjecture, attributed to Higman in the 1960s:

Higman's Conjecture. *For any Fuchsian group Γ, there exists $N(\Gamma)$ such that Γ surjects onto A_n for all $n > N(\Gamma)$.*

Conder's result covers the case $\Gamma = T_{2,3,7}$. Higman's conjecture was eventually proved in 2000 by Everitt [21]. One can quickly reduce to the case of genus 0— for example, a Fuchsian group of genus $g \geq 2$ maps onto the free group F_2 on generators x, y (send $a_1 \to x, a_2 \to y$ and all other generators to 1), and then F_2 maps onto A_n. For the genus 0 case, Everitt's approach was based on constructing generators using Higman's method of coset diagrams.

Inspired by Everitt's result, Liebeck and Shalev started thinking a few years later about whether there might be probabilistic approach to Higman's conjecture: if one could show that $P_\Gamma(A_n) = \text{Prob}(\phi \in \text{Hom}(\Gamma, A_n)$ is epi) tends to 1 (or indeed anything nonzero) as $n \to \infty$, then of course Higman's conjecture would follow.

Our approach to studying $P_\Gamma(A_n)$ was similar to the approach to Dixon's conjecture. Namely, if $\phi \in \text{Hom}(\Gamma, A_n)$ is not an epimorphism, then $\phi(\Gamma) \leq M$ for some maximal subgroup M of A_n; and given M, this happens with probability $|\text{Hom}(\Gamma, M)|/|\text{Hom}(\Gamma, A_n)|$. Therefore

$$1 - P_\Gamma(A_n) \leq \sum_{M \, max \, A_n} \frac{|\text{Hom}(\Gamma, M)|}{|\text{Hom}(\Gamma, A_n)|}.$$

Hence the task was to estimate $|\text{Hom}(\Gamma, A_n)|$ and also $|\text{Hom}(\Gamma, M)|$ for maximal subgroups M. Notice that these estimates must be delicate enough to distinguish between $|\text{Hom}(\Gamma, A_n)|$ and $|\text{Hom}(\Gamma, A_{n-1})|$, since A_{n-1} is one possibility for M.

Here is an example of an ingredient of how $|\text{Hom}(\Gamma, A_n)|$ is estimated, for the triangle group $\Gamma = T_{2,3,7}$, taken from [55]. For convenience take n to be divisible by $2 \cdot 3 \cdot 7$ and such that the conjugacy classes C_1, C_2, C_3 consisting of fixed-point-free permutations of shapes $(2^{n/2}), (3^{n/3}), (7^{n/7})$ respectively, all lie in A_n. One can prove that for $r = 2, 3, 7$ the following hold:

(i) $|C_r| \sim (n!)^{1-\frac{1}{r}}$.

(ii) For any $\chi \in \text{Irr}(A_n)$ and $c_r \in C_r$, we have $|\chi(c_r)| < cn^{1/2} \cdot \chi(1)^{1/r}$.

Part (i) is routine, but (ii) is hard and uses much of the well-developed character theory of symmetric groups. If we ignore the $cn^{1/2}$ term in (ii), then, writing $\mathbf{C} = (C_1, C_2, C_2)$, the formula (1.6) gives

$$|\mathrm{Hom}_C(\Gamma, A_n)| \geq \frac{(n!)^{\frac{1}{2}+\frac{2}{3}+\frac{6}{7}}}{n!}(1 - \sum_{1 \neq \chi \in Irr(A_n)} \frac{\chi(1)^{\frac{1}{2}+\frac{1}{3}+\frac{1}{7}}}{\chi(1)})$$

$$= (n!)^{\frac{43}{42}}(1 - (\zeta^{A_n}(\tfrac{1}{42}) - 1)).$$

Now $\mu(\Gamma) = \mu(T_{2,3,7}) = \frac{1}{42}$, and $\zeta^{A_n}(\frac{1}{42}) \to 1$ as $n \to \infty$ by Theorem 1.30(i), so for large n this shows that $|\mathrm{Hom}(\Gamma, A_n)|$ is at least roughly $(n!)^{1+\mu(\Gamma)}$. Adapting this calculation to take care of the $cn^{1/2}$ term in (ii) is a fairly routine technical matter.

It turns out that $(n!)^{1+\mu(\Gamma)}$ is the correct order of magnitude for $|\mathrm{Hom}(\Gamma, A_n)|$ and also $|\mathrm{Hom}(\Gamma, S_n)|$. The following result is [55, Theorem 1.2].

Theorem 1.31. *For any Fuchsian group Γ we have $|\mathrm{Hom}(\Gamma, S_n)| = (n!)^{1+\mu(\Gamma)+o(1)}$.*

In fact some much more precise estimates were obtained (and were needed, as remarked above). We were also able to prove that the subgroup growth function $a_n(\Gamma) = |\mathrm{Hom}_{\mathrm{trans}}(\Gamma, S_n)|/(n-1)!$ satisfies $a_n(\Gamma) = (n!)^{\mu(\Gamma)+o(1)}$, and establish the following probabilistic result ([55, Theorem 1.7]).

Theorem 1.32. *For any Fuchsian group Γ, the probability that a random homomorphism in $\mathrm{Hom}_{\mathrm{trans}}(\Gamma, S_n)$ is an epimorphism tends to 1 as $n \to \infty$.*

This implies Higman's conjecture, which was our original motivation. The character-theoretic methods and estimates in [55] for symmetric and alternating groups have been developed and improved in a number of subsequent papers, notably Larsen–Shalev [42], where very strong estimates on character values are obtained and a variety of applications given; in particular they solve a number of longstanding problems concerning mixing times of random walks on symmetric groups. But we shall not go into this here.

1.3.4 Groups of Lie Type

We now discuss results analogous to Theorems 1.31 and 1.32 for groups $G = G(q)$ of Lie type. Let Γ be a co-compact Fuchsian group as in Sect. 1.3.1. Again we seek to apply Lemma 1.29. If g_i $(1 \leq i \leq d)$ are elements of G of order dividing m_i, and $\chi \in \mathrm{Irr}(G)$, then $\frac{|\chi(g_1)\cdots\chi(g_d)|}{\chi(1)^{2g-2+d}} \leq \chi(1)^{-(2g-2)}$, which quickly yields

$$2 - \zeta^G(2g-2) \leq \sum_{\chi \in Irr(G)} \frac{\chi(g_1)\cdots\chi(g_d)}{\chi(1)^{2g-2+d}} \leq \zeta^G(2g-2).$$

This observation is only useful when $g \geq 2$, in which case Theorem 1.30 gives $\zeta^G(2g-2) \to 1$ as $|G| \to \infty$. In this case, applying Lemma 1.29 and summing over all classes of elements of orders dividing $m_1, \ldots, m_d$ leads to the following result, which is part of [57, 1.2, 1.4]. In the statement, $j_m(G)$ denotes the number of elements of G of order dividing m.

Theorem 1.33. *Let Γ be a co-compact Fuchsian group of genus $g \geq 2$ as in Sect. 1.3.1.*

(i) *For all finite quasisimple groups G,*

$$|\mathrm{Hom}(\Gamma, G)| = (1 + o(1)) \cdot |G|^{2g-1} \cdot \prod_{1}^{d} j_{m_i}(G),$$

where $o(1)$ refers to a quantity which tends to 0 as $|G| \to \infty$.

(ii) *For groups G of Lie type of rank r, $|\mathrm{Hom}(\Gamma, G)| = |G|^{1 + \mu(\Gamma) + O(\frac{1}{r})}$.*

Part (ii) follows from (i), since one can show that for groups G of large rank, $j_m(G)$ is roughly $|G|^{1 - \frac{1}{m}}$.

As in the previous section, the analysis of the probabilities $P_\Gamma(G)$ is much harder, and for groups of Lie type has only been completed for the case where $g \geq 2$. Here is [57, Theorem 1.6].

Theorem 1.34. *Let Γ be a co-compact Fuchsian group of genus $g \geq 2$. Then for all finite simple groups G, the probability $P_\Gamma(G) \to 1$ as $|G| \to \infty$.*

Note that the conclusion of the theorem does not remain true for genus 0 or 1, since there are Fuchsian groups of such genus which do not have all sufficiently large finite simple groups as quotients. For example, a result of Macbeath [68] says that $PSL_2(q)$ can only be an image of the genus 0 group $T_{2,3,7}$ if $q = p$ or p^3 for some prime p. Nevertheless, the genus 0 or 1 case does lead to some very interesting questions which we shall discuss below in Sect. 1.3.6. For the moment, we conclude this section by stating a conjecture from [57]:

Conjecture. *For any Fuchsian group Γ there is an integer $f(\Gamma)$ such that for finite simple classical groups G of rank at least $f(\Gamma)$, we have $P_\Gamma(G) \to 1$ as $|G| \to \infty$.*

1.3.5 Representation Varieties

Let Γ be a Fuchsian group, and let $K = \bar{\mathbb{F}}_p$, the algebraic closure of $\mathbb{F}_p$, where p is prime. Recall that the representation variety of Γ in dimension n over K is the variety $V = \mathrm{Hom}(\Gamma, GL_n(K))$. One of the most basic questions about V is: what is the dimension of V?

This question can be attacked using results in the previous section. Here's how. Let q be a power of the characteristic p, and define the field morphism $\sigma : GL_n(K) \to GL_n(K)$ to be the map sending the matrix (a_{ij}) to (a_{ij}^q). The fixed point group of σ is $GL_n(K)_\sigma = GL_n(q)$. Observe that σ also acts on the variety

V: for $\phi \in V, \gamma \in \Gamma$, define $\phi^\sigma(\gamma) = \phi(\gamma)^\sigma$. Then the fixed point set $V_\sigma = V(q)$ is the finite representation variety $\mathrm{Hom}(\Gamma, GL_n(q))$.

When the genus of Γ is at least 2, the size of $\mathrm{Hom}(\Gamma, GL_n(q))$ can be estimated as in the previous section. The connection between this and the dimension of V is given by a classical result of Lang–Weil [41]:

Lemma 1.35. *Let V be an algebraic variety over $\bar{\mathbb{F}}_p$ of dimension f, with e components of dimension f. For a power q of p, let $V(q)$ be the set of q-rational points in V. Then there is a power q_0 of p such that $|V(q)| = (e + o(1))q^f$ for all powers q of q_0.*

In estimating $|V(q)| = |\mathrm{Hom}(\Gamma, GL_n(q))|$ we need to know the limiting behaviour of the zeta function $\zeta^{GL_n(q)}(s)$. This is a little different from Theorem 1.30, because $GL_n(q)$ has $q - 1$ linear characters (which contribute $q - 1$ to $\zeta^{GL_n(q)}(s)$). The outcome is [57, 2.10]: for $s \geq 2$ and fixed n,

$$\zeta^{GL_n(q)}(s) \to q - 1 + \delta \text{ as } q \to \infty,$$

where $\delta = 1$ if $n = s = 2$ and $\delta = 0$ otherwise. Using this, $|\mathrm{Hom}(\Gamma, GL_n(q))|$ can be computed as in the previous section. The conclusion is a little awkward to state in general, so we just state it for surface groups and refer the reader to [57, 3.8] for the general case.

Proposition 1.36. *If Γ is a surface group of genus $g \geq 2$, then for fixed $n \geq 2$, $|\mathrm{Hom}(\Gamma, GL_n(q))| = (q - 1 + \delta + o(1)) \cdot |GL_n(q)|^{2g-1}$.*

By Lemma 1.35, this implies

Theorem 1.37. *Let Γ be a surface group of genus $g \geq 2$, and let V be the representation variety $\mathrm{Hom}(\Gamma, GL_n(K))$. Then $\dim V = (2g - 1)n^2 + 1$, and V has a unique irreducible component of highest dimension.*

Arguing similarly for the varieties $\mathrm{Hom}(\Gamma, \bar{G})$, where $\bar{G} = G(K)$ is a simple algebraic group over K, we obtain the following, which is [57, 1.10]. In the statement, $J_m(\bar{G})$ is the subvariety of elements $x \in \bar{G}$ satisfying $x^m = 1$. The dimensions of these subvarieties are studied by Lawther in [45]; in particular, $\dim J_m(\bar{G})$ tends to $(1 - \frac{1}{m}) \dim \bar{G}$ as the rank of $\bar{G}$ tends to infinity.

Theorem 1.38. *Let Γ be a Fuchsian group of genus $g \geq 2$ as in Sect. 1.3.1, and let $\bar{G} = G(K)$ be a simple algebraic group. If V is the variety $\mathrm{Hom}(\Gamma, \bar{G})$, then*

$$\dim V = (2g - 1) \dim \bar{G} + \sum_1^d \dim J_{m_i}(\bar{G}).$$

Other results along these lines can be found in [57].

1.3.6 Triangle Groups

All the results in the previous two sections concerning the spaces $\text{Hom}(\Gamma, G)$ for G a finite or algebraic group of Lie type assume that the genus of Γ is at least 2. This is not because the genus 0 or 1 cases are uninteresting, but rather because the character-theoretic methods in these cases require much more delicate information. For example, to estimate the sum in the formula (1.6), one needs information on the character values $\chi(g)$ for irreducible characters χ of G, and usable estimates of character values are hard to come by.

As a result, rather little is known about the spaces $\text{Hom}(\Gamma, G)$ and their ramifications when G is a group of Lie type and Γ has genus 0 or 1. (Note however that this is not so for $G = S_n$ or A_n, since the results of Sect. 1.3.3 hold for all genera.) Nevertheless there are some results and conjectures which make this a very interesting case.

We shall focus on the case for which most is known—namely that in which Γ is a triangle group $T_{a,b,c}$ (of genus 0).

Triangle Generation

Let $T = T_{a,b,c} = \langle x, y, z : x^a = y^b = z^c = xyz = 1 \rangle$ be a Fuchsian triangle group (so $\frac{1}{a} + \frac{1}{b} + \frac{1}{c} < 1$). We shall say that a finite group G is (a, b, c)-*generated* if it is an image of T. And for a family of simple groups $G(q)$ of fixed Lie type, we say that $G(q)$ is *randomly* (a, b, c)-*generated* if

$$P_T(G(q)) = \text{Prob}(\text{random } \phi \in \text{Hom}(T, G(q)) \text{ is epi}) \to 1$$

as $q \to \infty$ through values for which a, b and c divide $|G(q)|$.

Question. *Which finite simple groups of Lie type are (randomly) (a, b, c)-generated?*

This question goes back quite a long way, particularly in the case where $(a, b, c) = (2, 3, 7)$—indeed, $(2, 3, 7)$-generated groups are also called *Hurwitz groups*, and are of interest because they are precisely the groups which realize a well known upper bound of Hurwitz for the number of automorphisms of a compact Riemann surface (see [14] for example).

The first substantial result on this question was that of Macbeath [68], who showed that $PSL_2(q)$ is $(2, 3, 7)$-generated if and only if either $q = p \equiv 0, \pm 1 \bmod 7$, or $q = p^3$ and $p \equiv \pm 2, \pm 3 \bmod 7$, where p denotes a prime. Many further results on $(2, 3, 7)$-generation have followed since—for example, $PSL_3(q)$ is only $(2, 3, 7)$-generated if $q = 2$, while $SL_n(q)$ is $(2, 3, 7)$-generated for all $n \geq 287$ (see [87] for a survey).

Concerning (a, b, c)-generation for more general values, nothing much was done until the following two results of Marion [70, 72]. In both results, assume a, b, c are *primes* and $a \leq b \leq c$.

Theorem 1.39. *Given a prime p, there is a unique power p^r such that $PSL_2(p^r)$ is (a, b, c)-generated—namely, the minimal power such that a, b and c divide $|PSL_2(p^r)|$.*

A few moments' thought show that this agrees with Macbeath's result on $(2, 3, 7)$-generation stated above. In particular, the theorem shows that $PSL_2(q)$ is far from being randomly (a, b, c)-generated.

For three-dimensional classical groups Marion proved

Theorem 1.40. *Let $G = PSL_3(q)$ or $PSU_3(q)$.*

 (i) *If $a > 2$, then G is randomly (a, b, c)-generated.*
(ii) *If $a = 2$, then given a prime p, there are at most four values of $q = p^r$ such that G is (a, b, c)-generated.*

The proofs of these two theorems are character-theoretic. The second theorem takes a great deal of effort, and appears in a series of three papers [72]. Using the same methods to tackle higher dimensional groups is not an appetising prospect.

The dichotomy in parts (i) and (ii) of Theorem 1.40 is quite striking. In seeking to explain it, Marion introduced the idea of rigidity.

Rigidity

Now switch attention to $\bar{G}$, a simple algebraic group over $K = \bar{\mathbb{F}}_p$, p prime. For $a \geq 2$, define

$$\delta_a = \max\{\dim x^{\bar{G}} : x \in \bar{G} \text{ of order } a\}.$$

Straightforward matrix calculations give

Lemma 1.41. (i) *If $\bar{G} = PSL_2(K)$ then $\delta_a = 2$ for all $a \geq 2$.*
(ii) *If $\bar{G} = PSL_3(K)$ then $\delta_2 = 4$, while $\delta_a = 6$ for $a > 2$.*

For example, consider $G = PSL_3(K)$ with $\text{char}(K) \neq 2$. Any involution $t \in G$ is conjugate to the image modulo scalars of the diagonal matrix $\text{diag}(-1, -1, 1)$; then $\dim C_G(t) = 4$ and so $\dim t^G = 4$. On the other hand, for any odd prime $a \neq \text{char}(K)$ there is an element $u \in G$ of order a having distinct eigenvalues, so that $\dim C_G(u) = 2$ and $\dim u^G = 6$.

The relevance of the above definition to (a, b, c)-generation is given by

Proposition 1.42. *If $\delta_a + \delta_b + \delta_c < 2 \dim \bar{G}$, then $G(q)$ is not (a, b, c)-generated for any q.*

This is [71, Proposition 1]. The proof is an application of a well known result of Scott [82] which is one of the main tools in this whole area. Here is a sketch for the case where p is a "very good prime" for $\bar{G}$—this means that p is not 2 when $\bar{G}$ is symplectic or orthogonal, p is not 2 or 3 when $\bar{G}$ is of exceptional type (and also not 5 when $\bar{G} = E_8$), and p does not divide n when $\bar{G} = PSL_n(K)$.

We consider the action of $\bar{G}$ on its Lie algebra $V = L(\bar{G})$. Under the assumption that p is very good, $G(q)$ acts irreducibly on V for any q, and also $\dim C_V(g) = \dim C_{\bar{G}}(g)$ for all $g \in \bar{G}$ (see [12, 1.14]). If $G(q)$ is generated by x_1, x_2, x_3 of orders a, b, c with $x_1 x_2 x_3 = 1$, then Scott's result implies that $\sum (\dim V - \dim C_V(x_i)) \geq 2 \dim V$. As $\dim V = \dim \bar{G}$, this means that $\sum \dim x_i^{\bar{G}} \geq 2 \dim \bar{G}$, contradicting the hypothesis that $\delta_a + \delta_b + \delta_c < 2 \dim \bar{G}$.

The proposition motivates the following definition.

Definition. We say that a triple (a, b, c) of primes (with $\frac{1}{a} + \frac{1}{b} + \frac{1}{c} < 1$) is *rigid* for $\bar{G}$ if $\delta_a + \delta_b + \delta_c = 2 \dim \bar{G}$.

For example, if $\bar{G} = PSL_2(K)$ then $\dim \bar{G} = 3$, so by Lemma 1.41, every triple is rigid for $\bar{G}$. On the other hand, if $\bar{G} = PSL_3(K)$ then $\dim \bar{G} = 8$ and triples $(2, b, c)$ are rigid, while triples (a, b, c) with $a > 2$ are not (recall that $a \leq b \leq c$).

In view of these examples, Theorems 1.39 and 1.40 support the following conjecture, stated in [71].

Conjecture. *Let p be a prime and $\bar{G}$ a simple algebraic group over $K = \bar{\mathbb{F}}_p$. Suppose (a, b, c) is a rigid triple of primes for $\bar{G}$. Then there are only finitely many powers $q = p^r$ such that $G(q)$ is (a, b, c)-generated.*

Note that the converse to the conjecture does not hold: for example, it is known that $SL_7(q)$ is never a Hurwitz group (see [75]), but $(2, 3, 7)$ is not a rigid triple for $SL_7(K)$. It would be very interesting to find a variant of the conjecture which could work both ways round.

Theorem 3 of [71] classifies all the rigid triples of primes for simple algebraic groups. The list is not too daunting: for $\bar{G} = PSL_n$ or SL_n, rigid triples exist only for $n \leq 10$; for symplectic groups, only for dimensions up to 26 (and only the Hurwitz triple $(2, 3, 7)$ can be rigid beyond dimension 10); for orthogonal groups, the only rigid triple is $(2, 3, 7)$ for the groups $\mathrm{Spin}_{11,12}$; and the only exceptional type which has a rigid triple is G_2 with triple $(2, 5, 5)$.

As remarked above, one would not want to adopt the character-theoretic method of proof of Theorems 1.39 and 1.40 for larger rank cases. Fortunately there is another tool which can be used to attack the conjecture, namely the classical notion of rigidity for algebraic groups.

Classical Rigidity

Let $\bar{G}$ be a simple algebraic group over an algebraically closed field K, and let $C_1, \ldots, C_r$ be conjugacy classes in $\bar{G}$. Define

$$C_0 = \{(x_1, \ldots, x_r) : x_i \in C_i, \; x_1 \cdots x_r = 1\}.$$

Following [86], we say that $(C_1, \ldots, C_r)$ is a *rigid* tuple of classes if $C_0 \neq \emptyset$ and G acts transitively on C_0 by conjugation.

Rigid tuples play a major role in inverse Galois theory and other areas. Their relevance to triangle generation and in particular to the above Conjecture is via the following observations. Assume for convenience of discussion that the characteristic of K is a very good prime for $\bar{G}$.

1. If (C_1, C_2, C_3) is a rigid triple of classes, and $C_{L(\bar{G})}(x_1, x_2, x_3) = 0$ for $(x_1, x_2, x_3) \in C_0$, then $\sum \dim C_i = 2 \dim \bar{G}$ (see [86, 3.2]).
2. Suppose one can prove a converse to (1)—namely, that if (C_1, C_2, C_3) is a triple of classes such that $\sum \dim C_i = 2 \dim \bar{G}$ and $C_{L(\bar{G})}(x_1, x_2, x_3) = 0$ for $(x_1, x_2, x_3) \in C_0$, then (C_1, C_2, C_3) is a rigid triple.
3. Now let (a, b, c) be a rigid triple of primes. Simple algebraic groups have only finitely many classes of elements of any given order, and hence $\bar{G}$ has finitely many triples (C_1, C_2, C_3) of classes of elements of orders a, b, c satisfying $\dim C_i = 2 \dim \bar{G}$ (and the other triples of such classes have dimensions summing to less than $2 \dim \bar{G}$).

Given (1), (2) and (3), define $\mathcal{T}$ to be the set of triples (g_1, g_2, g_3) of elements of $\bar{G}$ of orders a, b, c with product 1 such that $C_{L(\bar{G})}(g_1, g_2, g_3) = 0$ and $\sum \dim C_i = 2 \dim \bar{G}$, where $C_i = g_i^{\bar{G}}$. For $(g_1, g_2, g_3) \in \mathcal{T}$, the triple of classes (C_1, C_2, C_3) is one of the finitely many in (3), and is rigid by (2). It follows that $\bar{G}$ has only finitely many orbits in its conjugation action on $\mathcal{T}$, and so there can be only finitely many groups $G(q)$ generated by such triples g_1, g_2, g_3, proving the Conjecture.

Hence, if we can prove the statement in (2), then we can prove the Conjecture. Unfortunately, the only known case of (2) is the following result of Strambach and Völklein [86, 2.3] for $\bar{G} = SL_n$; in the case of characteristic zero it goes back to Katz [39].

Theorem 1.43. *Let* $\bar{G} = SL_n(K)$, *and let* $(C_1, \ldots, C_r)$ *be a tuple of classes in* $\bar{G}$ *such that* $\sum \dim C_i = 2 \dim \bar{G}$ *and* $\langle x_1, \ldots, x_r \rangle$ *is irreducible on the natural module* $V_n(K)$ *for some* $(x_1, \ldots, x_r) \in C_0$. *Then* $(C_1, \ldots, C_r)$ *is a rigid tuple.*

The Conjecture follows in the case where $\bar{G} = SL_n$ (see [71] for details). Further cases are handled in [71], but quite a few remain open. The most elegant way to finish the proof would be to prove a version of Theorem 1.43 for all types of simple algebraic groups, but this seems difficult. One further case—that in which $\bar{G} = G_2$ in characteristic 5 and $(a, b, c) = (2, 5, 5)$ (the only rigid triple of primes for exceptional types)—was handled in [62].

Recently, Larsen et al. [43] have introduced the method of deformation theory into the picture, and used it to prove Marion's conjecture for all cases where the underlying characteristic does not divide the product abc of the exponents of the triangle group.

1.4 Cayley Graphs of Simple Groups: Diameter and Growth

Let G be a finite group with a generating set S which is symmetric—that is, closed under taking inverses—and does not contain the identity. The *Cayley graph* $\Gamma(G, S)$ is defined to be the graph with vertex set G and edges $\{g, gs\}$ for all $g \in G, s \in S$.

It is connected and regular of valency $|S|$, and G acts regularly on $\Gamma(G, S)$ by left multiplication. Because of the transitive action of G, the diameter of $\Gamma(G, S)$, denoted by $\mathrm{diam}(G, S)$, is equal to the maximum distance between the identity element and any $g \in G$, and so

$$\mathrm{diam}(G, S) = \max\{l(g) : g \in G\}$$

where $l(g)$ is the length of the shortest expression of g as a product of elements of S. If $d = \mathrm{diam}(G, S)$, then $G = \{e\} \cup \bigcup_{r=1}^{d} S^r$ (where $S^r = \{s_1 \cdots s_r : s_i \in S\}$), and so $|G| \leq \sum_{r=0}^{d} |S|^r < |S|^{d+1}$. Hence

$$\mathrm{diam}(G, S) > \frac{\log |G|}{\log |S|} - 1. \tag{1.7}$$

Examples. 1. Let $G = C_n = \langle x \rangle$, a cyclic group of order n, and let $S = \{x, x^{-1}\}$. Then $\Gamma(G, S)$ is an n-gon. So $\mathrm{diam}(G, S) = [\frac{n}{2}]$, whereas $\frac{\log |G|}{\log |S|} = \frac{\log n}{\log 2}$.

2. Let $G = S_n$ and S be the set of all transpositions. Here $\mathrm{diam}(G, S) = n - 1$, while $\frac{\log |G|}{\log |S|}$ is roughly $\frac{n}{2}$.

3. Let $G = S_n$ and $S = \{(1\,2), (1\,2 \cdots n)^{\pm 1}\}$. In this case $\mathrm{diam}(G, S)$ is roughly n^2, while $\frac{\log |G|}{\log |S|}$ is of the order of $n \log n$. The same orders of magnitude apply to a similar generating set for A_n consisting of a 3-cycle and an n- or $(n-1)$-cycle and their inverses.

4. For $G = SL_n(q)$ and S the set of transvections, we have $\mathrm{diam}(G, S) \sim n$ and $\frac{\log |G|}{\log |S|} \sim \frac{n}{2}$.

5. Let $G = SL_n(p)$ (p prime) and $S = \{x^{\pm 1}, y^{\pm 1}\}$ where

$$x = \begin{pmatrix} 1 & 1 & & & \\ & 1 & & & \\ & & \cdot & & \\ & & & \cdot & \\ & & & & \cdot \\ & & & & 1 \end{pmatrix}, \quad y = \begin{pmatrix} 0 & 1 & & & \\ 0 & 0 & 1 & & \\ & & & \cdot & \\ & & & & \cdot \\ & & & & 1 \\ \pm 1 & & & & \end{pmatrix}.$$

Then $\frac{\log |G|}{\log |S|} \sim n^2 \log p$, and also $\mathrm{diam}(G, S) \sim n^2 \log p$.

All the above examples are elementary except the last, where the fact that $\mathrm{diam}(G, S) \leq C n^2 \log p$ for some constant C is a result of Kassabov and Riley [38].

Define $\mathrm{diam}(G)$ to be the maximum of $\mathrm{diam}(G, S)$ over all generating sets S. The main conjecture in the field is due to Babai, and appears as Conjecture 1.7 in [6]:

Babai's Conjecture. *There is a constant c such that* $\mathrm{diam}(G) < (\log |G|)^c$ *for any non-abelian finite simple group G.*

It can be seen from Example 3 above that c must be at least 2 for the conjecture to hold.

There has been a great deal of recent progress on this conjecture, but before presenting some of this we shall discuss the special case where S is a union of conjugacy classes, which has various other connections.

1.4.1 Conjugacy Classes

In the case where the generating set S is a union of classes (which occurs in Examples 2 and 4 above), a strong form of Babai's conjecture holds:

Theorem 1.44. *There is a constant C such that for any non-abelian finite simple group G and any non-identity union S of conjugacy classes of G,*

$$\text{diam}(G, S) < C \frac{\log |G|}{\log |S|}.$$

Indeed, $G = S^k$ for all $k \geq C \frac{\log |G|}{\log |S|}$.

This is the main theorem of [54]. In view of (1.7), the diameter bound is best possible, apart from reduction of the constant C.

Consequences

First we point out an obvious consequence. If S is the set of involutions in a simple group G, then of course S is a union of classes, and it is not hard to prove that $|S| > c|G|^{1/2}$ (see [51, 4.2, 4.3]). Hence Theorem 1.44 implies that every element of every simple group is a product of k involutions, for some absolute constant k. The same holds for elements of any fixed order.

A more substantial consequence concerns *word maps* on simple groups. For a group G and a nontrivial word $w = w(x_1, \ldots, x_d)$ in the free group of rank d, define

$$w(G) = \{w(g_1, \ldots, g_d) : g_i \in G\},$$

the set of w-values in G. For example, if $w = [x_1, x_2]$ or x_1^k, then $w(G)$ is the set of commutators or kth powers in G. Clearly $w(G)$ is a union of classes of G.

Given w, it is possible to show that there is a constant $c = c(w) > 0$ depending only on w, such that $|w(G)| > |G|^c$ for all simple groups G such that $w(G) \neq 1$ (see [54, 8.2]). Hence Theorem 1.44 gives

Corollary 1.45. *For any nontrivial word w, there is a constant $c = c(w)$ such that $w(G)^c = G$ for every finite simple group G for which $w(G) \neq 1$.*

A result of Jones [36] ensures that $w(G) \neq 1$ provided G is sufficiently large (i.e. provided $|G| > f(w)$ where this depends only on w). Recent work of Larsen, Shalev and Tiep, culminating in [44], shows that the number $c(w)$ in the corollary can be replaced by 2. So for example, every element of every sufficiently large simple group is a product of two commutators or two kth powers, and so on.

Notice that for certain words w, it is definitely not possible to replace $c(w)$ by 1. For example, the kth power word map x_1^k is not surjective on any finite group of order not coprime to k (since it is not injective). To date, there are just a few instances of word maps which have been shown to be surjective on all simple groups. The first was the commutator word: it was proved in [61] that every element of every non-abelian finite simple group is a commutator, a result known as the Ore conjecture. Some further surjective word maps such as $x_1^p x_2^p$ (p prime) are produced in [26, 63]. One might conjecture that every non-power word map is surjective on sufficiently large simple groups, but this has recently been shown to be false in [35]: for example, the word map $(x, y) \rightarrow x^2[x^{-2}, y^{-1}]^2$ is non-surjective on $PSL_2(p^{2r+1})$ for all non-negative integers r and all odd primes p such that $p^2 \not\equiv 1 \bmod 16$ and $p^2 \not\equiv 0, 1 \bmod 5$.

This is one of a number of "width" questions about simple groups. Another is the conjecture proposed in [64], that if A is any subset of size at least 2 in a finite simple group G, then G is a product of N conjugates of A for some $N \leq c \log |G| / \log |A|$, where c is an absolute constant. This has been proved in some cases in [23, 60, 64].

1.4.2 Babai's Conjecture

There have been spectacular recent developments on Babai's conjecture, both for groups of Lie type and for alternating groups. We shall discuss these separately.

Groups of Lie Type

For a long time, even $SL_2(p)$ (p prime) was a mystery as far as proving Babai's conjecture was concerned. Probably the first small (symmetric) generating set one thinks of for this group is

$$S = \{ \begin{pmatrix} 1 & 1 \\ 0 & 1 \end{pmatrix}^{\pm 1}, \begin{pmatrix} 1 & 0 \\ 1 & 1 \end{pmatrix}^{\pm 1} \}.$$

Babai's conjecture asserts that $\mathrm{diam}(G, S) < (\log p)^c$ for these generators. Surely this must be easy?

In fact it is not at all easy, and was proved by the following beautiful but indirect method (see [65]). First observe that the matrices in S, when regarded as integer matrices, generate $SL_2(\mathbb{Z})$. Now let $\Gamma(p)$ denote the congruence subgroup which is the kernel of the natural map $SL_2(\mathbb{Z}) \to SL_2(p)$. If $\mathbb{H}$ is the upper half plane and $X(p)$ denotes the Riemann surface $\Gamma(p)\backslash\mathbb{H}$, denote by $\lambda_1(X(p))$ the smallest eigenvalue for the Laplacian on $X(p)$. A theorem of Selberg [83] gives $\lambda_1(X(p)) \geq \frac{3}{16}$ for all p, and this can be used to show that the Cayley graphs $\{\Gamma_p = \Gamma(SL_2(p), S) : p \text{ prime}\}$ have their second largest eigenvalues bounded away from the valency, and hence that they form a family of *expander graphs*. This means that there is an *expansion* constant $c > 0$, independent of p, such that for every set A consisting of fewer than half the total number of vertices in Γ_p, we have $|\delta A| > c|A|$, where δA is the boundary of A—that is, the set of vertices not in A which are joined to some vertex in A. From the expansion property it is easy to deduce that Γ_p has logarithmic diameter, so that $\mathrm{diam}(\Gamma(SL_2(p), S)) < c \log p$, a strong form of Babai's conjecture.

One can adopt essentially the same method for the generators

$$\left\{ \begin{pmatrix} 1 & 2 \\ 0 & 1 \end{pmatrix}^{\pm 1}, \begin{pmatrix} 1 & 0 \\ 2 & 1 \end{pmatrix}^{\pm 1} \right\}$$

of $SL_2(p)$, since, while these do not generate $SL_2(\mathbb{Z})$, they do generate a subgroup of finite index therein. But what if we replace the 2s in these generators with 3s? Then the matrices generate a subgroup of infinite index in $SL_2(\mathbb{Z})$, and the above method breaks down. This question became known as Lubotzky's 1-2-3 problem, and was not solved until the breakthrough achieved by Helfgott [30]:

Theorem 1.46. *Babai's conjecture holds for* $G = SL_2(p)$. *That is,* $\mathrm{diam}(SL_2(p))$ $< (\log p)^c$, *where c is an absolute constant.*

Helfgott deduced this from his key proposition: for any generating set S of $G = SL_2(p)$, either $|S^3| > |S|^{1+\epsilon}$, or $S^k = G$, where $\epsilon > 0$ and k do not depend on p. (Later it was observed that one can take $k = 3$ here.) The heart of his proof is to relate the growth of powers of subsets A of G with the growth of the corresponding set of scalars $B = \mathrm{tr}(A) = \{\mathrm{tr}(x) : x \in A\}$ in $\mathbb{F}_p$ under sums and products. By doing this he could tap into the theory of additive combinatorics, using results such as the following, taken from [9]: if B is a subset of $\mathbb{F}_p$ with $p^\delta < |B| < p^{1-\delta}$ for some $\delta > 0$, then $|B \cdot B| + |B + B| > |B|^{1+\epsilon}$, where $\epsilon > 0$ depends only on δ.

Following Helfgott's result, there was a tremendous surge of progress in this area. Many new families of expanders were constructed in [8]. Helfgott himself extended his result to SL_3 in [31], and this has now been proved for all groups of Lie type of bounded rank in [10, 81]. As a consequence, we have

Theorem 1.47. *If* $G = G(q)$ *is a simple group of Lie type of rank r, then* $\mathrm{diam}(G) < (\log|G|)^{c(r)}$ *where $c(r)$ depends only on r.*

Again, the theorem is proved via a growth statement: for any generating set S of $G(q)$, either $|S^3| > |S|^{1+\epsilon}$, or $S^3 = G$, where $\epsilon > 0$ depends only on r.

These results, and particularly their developments into the theory of expanders, have many wonderful and surprising applications. For a survey of these developments and some of the applications, see [67].

Finally, let us remark that Babai's conjecture remains open for groups of Lie type of unbounded rank.

Alternating Groups

For the alternating groups A_n, Babai's conjecture is that there is a constant C such that $\mathrm{diam}(A_n) < n^C$. Until very recently, the best bound for $\mathrm{diam}(A_n)$ was that obtained by Babai and Seress in [5], where it was proved that

$$\mathrm{diam}(A_n) < \exp((1 + o(1)) \cdot (n \log n)^{1/2}) = \exp((1 + o(1)) \cdot (\log |A_n|)^{1/2}).$$

Babai and Seress also obtained a bound of the same magnitude for the diameter of an arbitrary subgroup of S_n in [6]; this is best possible, as can be seen by constructing a cyclic subgroup generated by a permutation with many cycles of different prime lengths. Various other partial results appeared at regular intervals, such as that in [7], where it was shown that if the generating set S contains a permutation of degree at most $0.33n$, then $\mathrm{diam}(A_n, S)$ is polynomially bounded. But no real progress was made on Babai's conjecture until a recent breakthrough of Helfgott and Seress [32]:

Theorem 1.48. *We have* $\mathrm{diam}(A_n) \leq \exp(O((\log n)^4 \log \log n))$, *where the implied constant is absolute.*

This does not quite prove Babai's conjecture, but it does prove that $\mathrm{diam}(A_n)$ is "quasipolynomial" (where a quasipolynomial function $f(n)$ is one for which $\log f(n)$ is polynomial in $\log n$), which represents a big step forward. The same paper also gives a bound of the same magnitude for the diameter of any transitive subgroup of S_n.

References

1. M. Aschbacher, On the maximal subgroups of the finite classical groups. Invent. Math. **76**, 469–514 (1984)
2. M. Aschbacher, R. Guralnick, Some applications of the first cohomology group. J. Algebra **90**, 446–460 (1984)
3. M. Aschbacher, L. Scott, Maximal subgroups of finite groups. J. Algebra **92**, 44–80 (1985)
4. L. Babai, The probability of generating the symmetric group. J. Comb. Theor. Ser. A **52**, 148–153 (1989)
5. L. Babai, A. Seress, On the diameter of Cayley graphs of the symmetric group. J. Comb. Theor. Ser. A **49**, 175–179 (1988)
6. L. Babai, A. Seress, On the diameter of permutation groups. Eur. J. Comb. **13**, 231–243 (1992)

7. L. Babai, R. Beals, A. Seress, in *On the Diameter of the Symmetric Group: Polynomial Bounds*. Proceedings of the Fifteenth Annual ACM-SIAM Symposium on Discrete Algorithms (ACM, New York, 2004), pp. 1108–1112

8. J. Bourgain, A. Gamburd, Uniform expansion bounds for Cayley graphs of $SL_2(\mathbb{F}_p)$. Ann. Math. **167**, 625–642 (2008)

9. J. Bourgain, N. Katz, T. Tao, A sum-product estimate in finite fields, and applications. Geom. Funct. Anal. **14**, 27–57 (2004)

10. E. Breuillard, B. Green, T. Tao, Approximate subgroups of linear groups. Geom. Funct. Anal. **21**, 774–819 (2011)

11. T.C. Burness, M.W. Liebeck, A. Shalev, Generation and random generation: from simple groups to maximal subgroups, preprint, available at http://eprints.soton.ac.uk/151795.

12. R.W. Carter, *Finite Groups of Lie Type: Conjugacy Classes and Complex Characters*. Pure and Applied Mathematics (Wiley-Interscience, New York, 1985)

13. M.D.E. Conder, Generators for alternating and symmetric groups. J. Lond. Math. Soc. **22**, 75–86 (1980)

14. M.D.E. Conder, Hurwitz groups: a brief survey. Bull. Am. Math. Soc. **23**, 359–370 (1990)

15. F. Dalla Volta, A. Lucchini, Generation of almost simple groups. J. Algebra **178**, 194–223 (1995)

16. J.D. Dixon, The probability of generating the symmetric group. Math. Z. **110**, 199–205 (1969)

17. J.D. Dixon, Asymptotics of generating the symmetric and alternating groups. Electron. J. Comb. **12**, 1–5 (2005)

18. J.D. Dixon, B. Mortimer, *Permutation Groups*. Graduate Texts in Mathematics, vol. 163 (Springer, New York, 1996)

19. E.B. Dynkin, Semisimple subalgebras of semisimple Lie algebras. Trans. Am. Math. Soc. **6**, 111–244 (1957)

20. P. Erdös, P. Turán, On some problems of a statistical group-theory II. Acta Math. Acad. Sci. Hung. **18**, 151–163 (1967)

21. B. Everitt, Alternating quotients of Fuchsian groups. J. Algebra **223**, 457–476 (2000)

22. R.K. Fisher, The number of non-solvable sections in linear groups. J. Lond. Math. Soc. **9**, 80–86 (1974)

23. N. Gill, I. Short, L. Pyber, E. Szabó, On the product decomposition conjecture for finite simple groups, available at arXiv:1111.3497

24. R.M. Guralnick, W.M. Kantor, Probabilistic generation of finite simple groups. J. Algebra **234**, 743–792 (2000)

25. R.M. Guralnick, M. Larsen, P.H. Tiep, Representation growth in positive characteristic and conjugacy classes of maximal subgroups. Duke Math. J. **161**, 107–137 (2012)

26. R.M. Guralnick, G. Malle, Products of conjugacy classes and fixed point spaces. J. Am. Math. Soc. **25**, 77–121 (2012)

27. R.M. Guralnick, M.W. Liebeck, J. Saxl, A. Shalev, Random generation of finite simple groups. J. Algebra **219**, 345–355 (1999)

28. P. Hall, The Eulerian functions of a group. Q. J. Math. **7**, 134–151 (1936)

29. J. Häsä, Growth of cross-characteristic representations of finite quasisimple groups of Lie type, preprint, Imperial College London (2012)

30. H.A. Helfgott, Growth and generation in $SL_2(\mathbb{Z}/p\mathbb{Z})$. Ann. Math. **167**, 601–623 (2008)

31. H.A. Helfgott, Growth in $SL_3(\mathbb{Z}/p\mathbb{Z})$. J. Eur. Math. Soc. **13**, 761–851 (2011)

32. H.A. Helfgott, A. Seress, On the diameter of permutation groups, Ann. Math. (to appear), available at arXiv:1109.3550

33. B. Huppert, *Endliche Gruppen I*. Die Grundlehren der Mathematischen Wissenschaften, Band 134 (Springer, Berlin, 1967)

34. A. Jaikin-Zapirain, L. Pyber, Random generation of finite and profinite groups and group enumeration. Ann. Math. **173**, 769–814 (2011)

35. S. Jambor, M.W. Liebeck, E.A. O'Brien, Some word maps that are non-surjective on infinitely many finite simple groups, Bulletin London Math. Soc. (to appear), available at arXiv:1205.1952

36. G.A. Jones, Varieties and simple groups. J. Austral. Math. Soc. **17**, 163–173 (1974)
37. W.M. Kantor, A. Lubotzky, The probability of generating a finite classical group. Geom. Dedicata **36**, 67–87 (1990)
38. M. Kassabov, T.R. Riley, Diameters of Cayley graphs of Chevalley groups. Eur. J. Comb. **28**, 791–800 (2007)
39. N. Katz, *Rigid Local Systems* (Princeton University Press, Princeton, 1996)
40. P. Kleidman, M.W. Liebeck, *The Subgroup Structure of the Finite Classical Groups.* London Mathematical Society Lecture Note Series, vol. 129 (Cambridge University Press, Cambridge, 1990)
41. S. Lang, A. Weil, Number of points of varieties over finite fields. Am. J. Math. **76**, 819–827 (1954)
42. M. Larsen, A. Shalev, Characters of symmetric groups: sharp bounds and applications. Invent. Math. **174**, 645–687 (2008)
43. M. Larsen, A. Lubotzky, C. Marion, Deformation theory and finite simple quotients of triangle groups, preprint, Hebrew University, 2012
44. M. Larsen, A. Shalev, P.H. Tiep, Waring problem for finite simple groups. Ann. Math. **174**, 1885–1950 (2011)
45. R. Lawther, Elements of specified order in simple algebraic groups. Trans. Am. Math. Soc. **357**, 221–245 (2005)
46. M.W. Liebeck, G.M. Seitz, Maximal subgroups of exceptional groups of Lie type, finite and algebraic. Geom. Dedicata **36**, 353–387 (1990)
47. M.W. Liebeck, G.M. Seitz, On the subgroup structure of exceptional groups of Lie type. Trans. Am. Math. Soc. **350**, 3409–3482 (1998)
48. M.W. Liebeck, G.M. Seitz, On finite subgroups of exceptional algebraic groups. J. Reine Angew. Math. **515**, 25–72 (1999)
49. M.W. Liebeck, G.M. Seitz, The maximal subgroups of positive dimension in exceptional algebraic groups. Mem. Am. Math. Soc. **169**(802), 1–227 (2004)
50. M.W. Liebeck, A. Shalev, The probability of generating a finite simple group. Geom. Dedicata **56**, 103–113 (1995)
51. M.W. Liebeck, A. Shalev, Classical groups, probabilistic methods, and the $(2, 3)$-generation problem. Ann. Math. **144**, 77–125 (1996)
52. M.W. Liebeck, A. Shalev, Maximal subgroups of symmetric groups. J. Comb. Theor. Ser. A **75**, 341–352 (1996)
53. M.W. Liebeck, A. Shalev, Simple groups, probabilistic methods, and a conjecture of Kantor and Lubotzky. J. Algebra **184**, 31–57 (1996)
54. M.W. Liebeck, A. Shalev, Diameters of simple groups: sharp bounds and applications. Ann. Math. **154**, 383–406 (2001)
55. M.W. Liebeck, A. Shalev, Fuchsian groups, coverings of Riemann surfaces, subgroup growth, random quotients and random walks. J. Algebra **276**, 552–601 (2004)
56. M.W. Liebeck, A. Shalev, Character degrees and random walks in finite groups of Lie type. Proc. Lond. Math. Soc. **90**, 61–86 (2005)
57. M.W. Liebeck, A. Shalev, Fuchsian groups, finite simple groups and representation varieties. Invent. Math. **159**, 317–367 (2005)
58. M.W. Liebeck, B.M.S. Martin, A. Shalev, On conjugacy classes of maximal subgroups of finite simple groups, and a related zeta function. Duke Math. J. **128**, 541–557 (2005)
59. M.W. Liebeck, L. Pyber, A. Shalev, On a conjecture of G.E. Wall. J. Algebra **317**, 184–197 (2007)
60. M.W. Liebeck, N. Nikolov, A. Shalev, A conjecture on product decompositions in simple groups. Groups Geom. Dyn. **4**, 799–812 (2010)
61. M.W. Liebeck, E.A. O'Brien, A. Shalev, P.H. Tiep, The Ore conjecture. J. Eur. Math. Soc. **12**, 939–1008 (2010)
62. M.W. Liebeck, A.J. Litterick, C. Marion, A rigid triple of conjugacy classes in G_2. J. Group Theor. **14**, 31–35 (2011)

63. M.W. Liebeck, E.A. O'Brien, A. Shalev, P.H. Tiep, Products of squares in finite simple groups. Proc. Am. Math. Soc. **140**, 21–33 (2012)
64. M.W. Liebeck, N. Nikolov, A. Shalev, Product decompositions in finite simple groups. Bull. Lond. Math. Soc. (to appear), available at arXiv:1107.1528
65. A. Lubotzky, *Discrete Groups, Expanding Graphs and Invariant Measures*. Progress in Mathematics, vol. 125 (Birkhäuser, Basel, 1994)
66. A. Lubotzky, The expected number of random elements to generate a finite group. J. Algebra **257**, 452–459 (2002)
67. A. Lubotzky, Expander Graphs in Pure and Applied Mathematics, available at arXiv:1105.2389
68. A.M. Macbeath, Generators of the linear fractional groups, in *Proceedings of the Symposium on Pure Math.*, vol. XII (American Mathematical Society, Providence, Rhode Island, 1969), pp. 14–32
69. A. Mann, Positively finitely generated groups. Forum Math. **8**, 429–459 (1996)
70. C. Marion, Triangle groups and $PSL_2(q)$. J. Group Theor. **12**, 689–708 (2009)
71. C. Marion, On triangle generation of finite groups of Lie type. J. Group Theor. **13**, 619–648 (2010)
72. C. Marion, Random and deterministic triangle generation of three-dimensional classical groups I-III. Comm. Algebra (to appear)
73. A. Maroti, M.C. Tamburini, Bounds for the probability of generating the symmetric and alternating groups. Arch. Math. **96**, 115–121 (2011)
74. B.M.S. Martin, Reductive subgroups of reductive groups in nonzero characteristic. J. Algebra **262**, 265–286 (2003)
75. L. Di Martino, M.C. Tamburini, A.E. Zalesskii, On Hurwitz groups of low rank. Comm. Algebra **28**, 5383–5404 (2000)
76. N.E. Menezes, M.R. Quick, C.M. Roney-Dougal, The probability of generating a finite simple group, Israel J. Math. (to appear)
77. G.A. Miller, On the groups generated by two operators. Bull. Am. Math. Soc. **7**, 424–426 (1901)
78. E. Netto, *The Theory of Substitutions and Its Applications to Algebra*, 2nd edn. (Chelsea Publishing Co., New York, 1964) (first published in 1892)
79. I. Pak, On probability of generating a finite group, preprint, http://www.math.ucla.edu/pak.
80. I. Pak, What do we know about the product replacement algorithm? in *Groups and Computation, III*, vol. 8, Columbus, OH, 1999 (Ohio State University Mathematical Research Institute Publications/de Gruyter, Berlin, 2001), pp. 301–347
81. L. Pyber, E. Szabó, Growth in finite simple groups of Lie type of bounded rank, available at arXiv:1005.1858
82. L.L. Scott, Matrices and cohomology. Ann. Math. **105**, 473–492 (1977)
83. A. Selberg, On the estimation of Fourier coefficients of modular forms. Proc. Symp. Pure Math. **8**, 1–15 (1965)
84. R. Steinberg, Generators for simple groups. Can. J. Math. **14**, 277–283 (1962)
85. R. Steinberg, Endomorphisms of linear algebraic groups. Mem. Am. Math. Soc. **80**, 1–108 (1968)
86. K. Strambach, H. Völklein, On linearly rigid tuples. J. Reine Angew. Math. **510**, 57–62 (1999)
87. M.C. Tamburini, M. Vsemirnov, Hurwitz groups and Hurwitz generation, in *Handbook of Algebra*, vol. 4 (Elsevier/North-Holland, Amsterdam, 2006), pp. 385–426
88. M.C. Tamburini, J.S. Wilson, N. Gavioli, On the $(2, 3)$-generation of some classical groups I. J. Algebra **168**, 353–370 (1994)
89. J. Whiston, Maximal independent generating sets of the symmetric group. J. Algebra **232**, 255–268 (2000)

Chapter 2
Estimation Problems and Randomised Group Algorithms

Alice C. Niemeyer, Cheryl E. Praeger, and Ákos Seress

2.1 Estimation and Randomization

2.1.1 Computation with Permutation Groups

In 1973, Charles Sims [89] proved the existence of the Lyons–Sims sporadic simple group Ly by constructing its action as a group of permutations of a set of cardinality 8,835,156 on a computer which could not even store and multiply the two generators of Ly in this smallest degree permutation representation for the group! The existence of this finite simple group, together with many of its properties, had been predicted by Richard Lyons [60], but proof of existence was not established until Sims' construction. Leading up to this seminal achievement, Sims [88] had developed concepts and computational methods that laid the foundation for his general theory of permutation group computation.

A.C. Niemeyer (✉)
Centre for the Mathematics of Symmetry and Computation, School of Mathematics and Statistics, The University of Western Australia, 35 Stirling Highway, Crawley, WA 6009, Australia
e-mail: alice.niemeyer@uwa.edu.au

C.E. Praeger
Centre for the Mathematics of Symmetry and Computation, School of Mathematics and Statistics, The University of Western Australia, 35 Stirling Highway, Crawley, WA 6009, Australia

King Abdulaziz University, Jeddah, Saudi Arabia
e-mail: cheryl.praeger@uwa.edu.au

Á. Seress
Centre for the Mathematics of Symmetry and Computation, School of Mathematics and Statistics, The University of Western Australia, 35 Stirling Highway, Crawley, WA 6009, Australia

The Ohio State University, Columbus, OH, USA
e-mail: akos@math.ohio-state.edu

A. Detinko et al. (eds.), *Probabilistic Group Theory, Combinatorics, and Computing*,
Lecture Notes in Mathematics 2070, DOI 10.1007/978-1-4471-4814-2_2,
© Springer-Verlag London 2013

Sims introduced the critical concept of a *base* of a permutation group G on a finite set Ω, namely a sequence of points $\alpha_1, \ldots, \alpha_b$ of Ω such that only the identity of G fixes all of them. For example, the dihedral group $D_{2n} = \langle a, b \rangle$ acting on $\{1, 2, \ldots, n\}$, where $a = (1, 2, \ldots, n)$ and $b = (2, n)(3, n-1) \ldots$, has a base $B = (1, 2)$, since only the identity of D_{2n} fixes both 1 and 2. Moreover the $2n$ elements $g \in D_{2n}$ produce $2n$ distinct image pairs $(1^g, 2^g)$ of the base B—for example, a maps B to $(2, 3)$, b maps B to $(1, n)$.

Sims observed that elements of a permutation group G could always be represented uniquely by the sequence of images of the points of a given base B. He exploited this potentially compact representation of group elements, ingeniously showing how to compute in G with these base images, via a so-called strong generating set of G relative to B. Sims' algorithm to construct a base and strong generating set, called the *Schreier–Sims algorithm*, is of fundamental importance for permutation group computation.

For groups possessing a small base, the Schreier–Sims algorithm is extremely efficient, but for some groups every base has size close to the cardinality $n = |\Omega|$ of the point set. For such groups, the methods are not effective. Examples of such large-base groups include the "giants": the alternating group $\mathrm{Alt}(\Omega) = A_n$ and the symmetric group $\mathrm{Sym}(\Omega) = S_n$, which have minimum-sized bases $(1, 2, \ldots, n-2)$ and $(1, 2, \ldots, n-1)$ respectively.

2.1.2 Recognising the Permutation Group Giants

For computational purposes, a finite permutation group G on Ω is given by a (usually small) set X of generators. The group G consists of all products of arbitrary length of elements from X. Since the Schreier–Sims algorithm is ineffective for computation with the giants $\mathrm{Alt}(\Omega)$ and $\mathrm{Sym}(\Omega)$, it is important to determine in advance (that is, before trying to find a base and strong generating set) whether or not a given permutation group $G = \langle X \rangle$ is one of these giants. Thus the question of identifying the giants $\mathrm{Alt}(\Omega)$ and $\mathrm{Sym}(\Omega)$, given only a generating set of permutations, was a central issue in the development of general purpose group theory computer systems.

Theoretically the problem of detecting these giants had engaged mathematicians from the earliest studies of group theory. Since the seminal work of Camille Jordan in the 1870s, it has been known that there are many kinds of permutations such that the only transitive permutation groups containing them are the giants (we say that $G \leq \mathrm{Sym}(\Omega)$ is *transitive* if each pair of points of Ω can be mapped one to the other by an element of G). The most beautiful of these results that identifies a large family of such elements is Jordan's theorem below.

Let us call a permutation $g \in S_n$ a *Jordan element* if g contains a p-cycle, for some prime p with $n/2 < p < n - 2$. For example, $g = (1, 2, 3, 4, 5)(6, 7) \in S_9$ is a Jordan element (with $n = 9$, $p = 5$).

Theorem 2.1. *If a transitive permutation group $G \leq S_n$ contains a Jordan element then G is A_n or S_n.*

Given a set of generators for $G \leq \text{Sym}(\Omega)$, it is easy to test whether G is transitive. Hence, recognising the giants boils down to the question: how prevalent are the Jordan elements in the giants? For a fixed prime $p \in (n/2, n-2)$, the number of elements in S_n containing a p-cycle is

$$\binom{n}{p}(p-1)!(n-p)! = \frac{n!}{p} \text{ (and } \frac{n!}{2p} \text{ in } A_n),$$

so the proportion of Jordan elements in A_n or S_n for this prime p is $1/p$, and therefore the proportion of Jordan elements in A_n or S_n is $\sum_{n/2<p<n-2} \frac{1}{p} \geq \frac{c}{\log n}$ for some constant c. For $n \geq 100$, c can be taken to be $1/5$, which follows by applying an inequality by Dusart [25, p. 414] to determine the number of primes p with $n/2 < p < n - 2$. So roughly c out of every $\log n$ independent, uniformly distributed random elements from S_n or A_n will be Jordan. That is to say, we should find a Jordan element with high probability by randomly selecting elements in a giant.

2.1.3 Monte Carlo Algorithms

How do we turn the comments above into a justifiable algorithm? We want to make some multiple of $\log n$ random selections from a transitive group G on n points which we suspect may be S_n or A_n, but as yet we have no proof of this fact. We hope, and expect, to find a Jordan element, thereby uncovering the secret and proving that G really is a giant S_n or A_n.

Formally, we model this process as a *Monte Carlo algorithm*. The Monte Carlo method was invented by Stanislaw Ulam in the 1940s; it was named after Monte Carlo Casino in Monaco which Ulam's uncle visited often (see the account in [62]). The characteristics of a Monte Carlo algorithm are that it completes quickly, but allows a small (user-controlled) probability of "error", that is, of returning an incorrect result. In our context, for a Monte Carlo algorithm, we begin with a prescribed bound on the error probability $\varepsilon \in (0, 1)$. The algorithm typically makes a number $N = N(\varepsilon)$ of random selections, depending on ε, this number being determined in advance to guarantee that the probability of an incorrect result is at most ε.

Here is a worked example of a Monte Carlo algorithm to recognise the giants S_n and A_n among transitive permutation groups on n points.

Algorithm 1: JORDAN

Input: A transitive subgroup $G = \langle x_1, \ldots, x_k \rangle \leq S_n$ and a real number $\varepsilon \in (0, 1)$ (the error
    probability bound);
Output: true or false;
# We hope the algorithm returns true if G is S_n or A_n – see the comments below;
for *up to* $N = \lceil (\log \varepsilon^{-1})(\log n)/c \rceil$ *random elements g from G* **do**
    if *g is a Jordan element* **then**
        return true;
    end
end
Return false;

Comments on the algorithm

1. The procedure completes after at most N repeats of the **if** statement, so it *is* an
   algorithm! If it returns true then $G = A_n$ or S_n by Jordan's Theorem 2.1. On the
   other hand, if the algorithm returns false then this may be incorrect, but only if
   G does equal A_n or S_n, and we failed to find a Jordan element.
2. We have
   Prob(we do not find a Jordan element, given that $G = A_n$ or $G = S_n$) $\leq$
   $\left(1 - \frac{c}{\log n}\right)^N < \varepsilon$.
   So Algorithm 1 is a Monte Carlo algorithm with error probability less than ε.
   This is a special kind of Monte Carlo algorithm: the result true is always correct,
   and the possibility of an incorrect result is confined to the case where false is
   returned. Such algorithms are called *one-sided Monte Carlo algorithms*.
3. This probability estimate assumes that the random selections made are independent and uniformly distributed. There are algorithms available for producing "approximately random" elements from a group given by a generating set; see [3, 18,
   24]. We shall not discuss the theoretical details of these algorithms or their practical performance. Rather we assume in our discussion of randomised algorithms
   that we are dealing with independent uniformly distributed random elements.
4. The design and discussion of this simple algorithm used concepts and results
   from group theory to prove correctness, and from number theory to establish the
   bound on the error probability. It is typical to gather and develop methods from
   a variety of mathematical areas to achieve good algorithm design and analysis.
5. Algorithm 1 is essentially the algorithm used in GAP [37] and MAGMA [15]
   for testing if a permutation group is a giant. It was first described by Parker and
   Nikolai [79], preceding Sims' work by a decade. The second author (Praeger)
   recalls numerous discussions with John Cannon, over a number of years, about
   the implementation of this algorithm in connection with his development of the
   computer algebra system CAYLEY (a precursor to MAGMA). There was much
   to learn about improving the practical performance of the algorithm to avoid its
   becoming a bottle-neck for permutation group computation. A wider class of
   "good elements" than the Jordan elements was used, based on generalisations of

Jordan's Theorem (see [94, 13.10] and [83]), and better methods were developed to produce "approximately random" elements.

2.1.4 What Kinds of Estimates and in What Groups?

Notice the role estimates played in Algorithm 1:

a lower bound for the proportion of Jordan elements gives an upper bound on the error probability.

Does it matter if the estimate is far from the true value? We might, for different reasons, propose one of two different answers to this question:

1. We might say "no", because if there are more Jordan elements than our estimates predict, then we simply find one more quickly and the algorithm confirms that "G is a giant" more efficiently.
2. We might say "yes", because if G is not a giant then we force the algorithm to do needless work in testing too large a number of random elements so that the algorithm runs more slowly than necessary on non-giants. Note that the algorithm will never find a Jordan element in a non-giant by Theorem 2.1, so the full quota of random elements will be tested before completion.

For general purpose algorithms such as Algorithm 1, which are used frequently on arbitrary permutation groups, the quality of the estimates really does matter. We should try to make estimates as good as possible, especially when they are used to analyze a time-critical module of a randomised algorithm.

In the computer algebra systems GAP and MAGMA, new algorithms are under development for computation with matrix groups and permutation groups. These employ a tree-like data structure which allows a "divide and conquer" approach, reducing to computations in normal subgroups and quotient groups. This approach (see Sect. 2.4.1) reduces many computational problems to the case of finite simple groups. Accordingly many of the topics chosen in this chapter are of relevance to computing with finite simple groups.

2.1.5 What Group is That: Recognising Classical Groups as Matrix Groups

As a more substantial example for group recognition, we describe an algorithm to recognise a finite classical group in its natural representation. By this, we mean that the algorithm will return the "name" of the group. We give a broad-brush description of the classical recognition algorithm developed in [72] generalising the Neumann–Praeger SL-recognition algorithm in [69].

The algorithm takes as input a subgroup G of a finite n-dimensional classical group $\text{Class}(n, q)$ over a finite field $\mathbb{F}_q$ of order q, such as the general linear group $\text{GL}(n, q)$ or a symplectic group $\text{Sp}(n, q)$, in its natural representation as a group of matrices acting on the underlying vector space $V(n, q)$. The subgroup G is given by a generating set of $n \times n$ matrices over $\mathbb{F}_q$. The algorithm seeks so-called *ppd elements* in G which we describe as follows.

For an integer $e > 1$, a *primitive prime divisor (ppd)* of $q^e - 1$ is a prime r dividing $q^e - 1$ such that r does not divide $q^i - 1$ for any $i < e$. It has been known for a long time that primitive prime divisors exist unless $q = 2$, $e = 6$, or $e = 2$ and $q + 1$ is a power of 2; see [97]. Superficially, primitive prime divisors seem interesting because the order of the classical group has the form

$$|\operatorname{Class}(n, q)| = q^{\text{some power}} \prod_{\text{various } i} (q^i \pm 1).$$

We define a *ppd-$(q; e)$ element* $g \in \operatorname{Class}(n, q)$ as an element with order divisible by a ppd of $q^e - 1$. The algorithm in [72] seeks two ppd elements, namely a ppd-$(q; e_1)$ and a ppd-$(q; e_2)$ element for $e_1 \neq e_2$ and $e_1, e_2 > n/2$, which satisfy various additional minor conditions described in [72, Sects. 2 and 9]. We call such a pair a *good ppd matrix pair*. Their importance lies in the following deep theorem [72, Theorem 4.8], the proof of which relies heavily on the finite simple group classification.

Theorem 2.2. *If $G \leq \operatorname{Class}(n, q)$ is irreducible on $V(n, q)$ and G contains a good ppd matrix pair, then (essentially) $G = \operatorname{Class}(n, q)$ or G is known explicitly.*

Thus, provided that (a) we can test efficiently whether G is irreducible on $V(n, q)$, (b) good ppd matrix pairs are sufficiently prevalent in $\operatorname{Class}(n, q)$ and are easily identifiable, and (c) the exceptions in Theorem 2.2 are easy to deal with, the good ppd matrix pairs could play the role of the Jordan elements used to identify the permutation group giants in Algorithm 1. We would then have an analogue of Algorithm 1 for classical groups, underpinned by considerably deeper theory than Jordan's Theorem 2.1. It would look like this:

Algorithm 2: RECOGNISECLASSICAL

Input: An irreducible subgroup $G = \langle X_1, \ldots, X_k \rangle \leq \operatorname{Class}(n, q)$ and a real number
$\quad \varepsilon \in (0, 1)$ (the error probability bound).
Output: true or false
\# If the output is true, we are certain that $G = \operatorname{Class}(n, q)$;
\# the output false may be incorrect;
for Many(*depending on n, q, ε) random elements $g \in G$* **do**
$\quad$ determine if g is a ppd element with the additional properties;
$\quad$ **if** *a good ppd matrix pair is found* **then**
$\quad\quad$ **if** *G is one of the exceptions* **then**
$\quad\quad\quad$ **return** false
$\quad\quad$ **else**
$\quad\quad\quad$ **return** true;
$\quad\quad$ **end**
$\quad$ **end**
end
return false;

Comments on the algorithm

1. Note that if Algorithm 2 returns true then G really is Class(n, q) by Theorem 2.2; while if it returns false then the result may be incorrect (namely if $G = $ Class(n, q) and we fail to find the good ppd matrix pair).
2. The missing ingredient is our knowledge of the presence of good ppd matrix pairs in Class(n, q), and an estimate of their proportion. We need a positive lower bound on their proportion in order to decide how Many random elements to test to ensure an error probability of at most ε. This is necessary to prove that we have a one-sided Monte Carlo algorithm.

Estimating the proportion of ppd-$(q; e)$ elements in Class(n, q): For the details involved in dealing with the additional properties we refer the reader to [72]. For $G = $ Class(n, q) and $e > n/2$, let ppd(G, e) be the proportion of ppd-$(q; e)$ elements in G. We give a few details for the general linear case.

Lemma 2.3. *Let $G = $ GL(n, q) and let $\frac{n}{2} < e \leq n$. Then $\frac{1}{e+1} \leq ppd(G, e) \leq \frac{1}{e}$.*

Proof. Let $g \in G$ be a ppd-$(q; e)$ element and let r be a ppd of $q^e - 1$ dividing $|g|$. By considering a power of g of order r, we can show that g leaves invariant a unique e-dimensional subspace U of $V(n, q)$, and acts irreducibly on U. Moreover the induced element $g|_U$ is a ppd-$(q; e)$ element in GL(U), and a straightforward counting argument (see [72, Lemma 5.1]) shows that ppd$(G, e) = $ ppd$($GL$(U), e)$.

In other words, we may assume that $n = e$ in the proof. With this assumption, we have g irreducible on $V(n, q)$, and each such element lies in a Singer cycle $S = Z_{q^e - 1}$ of G. All Singer cycles are conjugate in G, and distinct Singer cycles contain disjoint sets of irreducible elements. Moreover the number of Singer cycles is $|G : N_G(S)| = |G|/(e(q^e - 1))$ (see [69, Lemma 2.1]). Hence ppd(G, e) is equal to $(1/e) \times$ (the proportion of such elements in the cyclic group S).

This immediately gives ppd$(G, e) \leq 1/e$. We need one more observation to obtain the lower bound. Certainly each element of S of order not divisible by r lies in the unique subgroup S_0 of S of index r. Thus each element of $S \setminus S_0$ has order divisible by r, and hence ppd$(G, e) \geq (1/e) \times (1 - 1/r)$. Now e is the least positive integer such that $q^e \equiv 1 \pmod{r}$, and so q has order e modulo the prime r. This implies that e divides $r - 1$, and in particular $r \geq e + 1$. Hence ppd$(G, e) \geq (1/e) \times e/(e + 1) = 1/(e + 1)$. $\square$

A similar argument in [72, Theorem 5.7] shows that the bounds of Lemma 2.3 hold for the other classical groups for almost all values of e. Since each ppd element corresponds to just one e-value (because $e > n/2$), we can find a lower bound for the proportion of ppd elements in G by adding the lower bounds for ppd(G, e) over all relevant e. For GL(n, q), this is $\sum_{n/2 < e \leq n} 1/e \sim \log 2$ by Lemma 2.3. For the other classical groups, the values of e occurring all have the same parity (odd for unitary groups and even for symplectic and orthogonal groups), and for these groups the proportion of ppd elements is roughly $(\log 2)/2$ [72, Theorem 6.1].

These lower bounds (or rather, the equivalent ones we obtain in [72] after taking into account the additional conditions on the ppd elements) allow us to decide how many random selections to make in order to find a good ppd matrix pair with probability at least $1 - 1/\varepsilon$, and hence to determine the value for Many in Algorithm 2.

2.1.6 What Group is That: Recognising Lie Type Groups in Arbitrary Representations

Of course, we do not only encounter the classical groups in their natural representation. If G is a simple group of Lie type, given in any permutation or matrix group representation, and the characteristic p of G is known, then we may proceed by an extension of Algorithm 2. The procedure that we sketch was developed in [6].

Let e_1 and e_2 be the two highest *ppd exponents*, that is, integers e such that G contains elements of order divisible by a primitive prime divisor of $p^e - 1$. It was shown in [6] that for each pair of integers (e_1, e_2), there are at most seven isomorphism types of Lie type groups of characteristic p with e_1, e_2 as the highest ppd exponents in the group. Also, ppd elements with ppd exponents e_1 and e_2 are frequent enough that we encounter them in a random sample of size polynomial in the input length.

To distinguish between the possibilities for G with the same values e_1 and e_2, we consider the third highest ppd exponent in G and elements whose order is divisible by a product of two ppd primes, corresponding to certain chosen ppd exponents. The result is a polynomial-time Monte Carlo algorithm that names the isomorphism type of G, with one exception: a polynomial-size random sample may not distinguish the groups $Sp(2m, p^f)$ and $O(2m + 1, p^f)$, for odd primes p. This last ambiguity was handled by Altseimer and Borovik [1].

2.2 Proportions of Elements in Symmetric Groups

2.2.1 Notation

In this section we fix a set Ω and consider the symmetric group $Sym(\Omega)$ on Ω. When $\Omega = \{1, \ldots, n\}$ for some positive integer n we write S_n instead of $Sym(\{1, \ldots, n\})$. Elements of S_n are written in disjoint cycle notation. The *number of cycles* of a given element $g \in S_n$ denotes the number of disjoint cycles g has on $\{1, \ldots, n\}$ including fixed points.

2.2.2 Historical Notes

The study of proportions of permutations has been of interest to mathematicians for a long time. For example, in 1708 Monmort introduced and analyzed a game of 13 cards which he called "jeu de Treize" (the game of thirteen) in his book on the theory of games [64, pp. 54–64]. He later generalised the game to any number of cards numbered from 1 to n [65, pp. 130–143]. In the game, a player has n turns, each time announcing out loud the number of the turn and picking a card at random from the deck of n cards without replacing it. The game is won if each time the number of the card and the number announced are different. Leonhard Euler in *Solutio Quaestionis curiosae ex doctrina combinationum* [34] describes the game as follows: *Data serie quotcunque litterarum a, b, c, d, e etc., quarum numerus sit n, invenire quot modis earum ordo immutari possit, ut nullo in eo loco reperiatur, quem initio occupaverat.* This can be translated as *Given an arbitrary series (sequence) of letters a, b, c, d, e, ..., let the number of which be n, find in how many ways their order may be changed so that none reappears in the same place which it originally occupied.*[1] In [33] Euler showed that the number of solutions is the integer closest to $n!/e$. Earlier solutions had already been given; for example, Monmort presented a solution by Nicolas Bernoulli [65, pp. 301–302]. De Moivre also mentions the game already in the first edition of [23], and gives a solution in [23, Problem 35].

Today this problem is often called the hat-swapping problem: *Suppose n men each put a hat on a hat rack in a restaurant. When they leave they each choose a random hat. What is the probability that no man chooses his own hat?*

Nowadays we call a permutation in S_n which has no fixed points on $\{1, \ldots, n\}$ a *derangement*, and we would rephrase the game of thirteen, Euler's question or the hat-swapping problem as: How many derangements are there in S_n?

In this section we will focus on certain other proportions of elements in S_n. The proportions that we focus on arise either from algorithmic applications for permutation groups or from applications to classical groups of Lie type (see Sect. 2.3.2).

2.2.3 Orders of Permutations

The order of a permutation can easily be read off from its disjoint cycle notation; namely it is the least common multiple of the cycle lengths. One of the oldest results on the order of an element in a symmetric group is due to Landau, who determined how large the order of an element in S_n can be asymptotically.

Theorem 2.4 (Landau [51]).

$$\lim_{n \to \infty} \frac{\log \left(\max_{g \in S_n} (\mathrm{ord}(g)) \right)}{\sqrt{n \log(n)}} = 1.$$

---

[1]Translation by Peter M. Neumann, The Queen's College, University of Oxford.

Although the order of an element in S_n can be as large as the previous theorem suggests, Erdős and Turán were able to prove, in the first of a series of papers [27–32] on the subject of the statistics of permutations, that most elements have much smaller order.

Theorem 2.5 (Erdős and Turán [27]). *For $\varepsilon, \delta > 0$ there is a number $N_0(\varepsilon, \delta)$ such that for all $n \geq N_0(\varepsilon, \delta)$,*

$$\frac{|\{g \in S_n \mid (1/2 - \varepsilon) \log^2(n) \leq \log(\mathrm{ord}(g)) \leq (1/2 + \varepsilon) \log^2(n)\}|}{n!} \geq 1 - \delta.$$

Erdős and Turán proved many more insightful results on the order of elements in symmetric groups. For example, in [28] they investigated prime divisors of the order of elements in symmetric groups. In [29] they described for any x the limiting behaviour as n tends to infinity of the proportion of elements g in S_n for which $\log(\mathrm{ord}(g)) \leq \frac{1}{2} \log^2(n) + x \log^{3/2}(n)$. In [30] they considered, among other problems, the number of different values that $\mathrm{ord}(g)$ can have as g ranges over the elements of S_n.

Goh and Schmutz [39] prove that the logarithm of the average order of a random permutation in S_n is $c \sqrt{n/\log(n)}$, where $c = 2\sqrt{2 \int_0^\infty \log\log\left(\frac{e}{1-e^{-t}}\right) dt}$. This constant is approximately 2.99.

2.2.4 Number of Cycles

Let $a(n)$ denote the average number of cycles of the elements in S_n. In a seminal paper [40], Gončarov examined various properties of random permutations. Among many other results, he proved that the average number of cycles of a permutation in S_n is close to $\log(n)$.

Theorem 2.6 (Gončarov [40]).

$$a(n) = \sum_{i=1}^{n} \frac{1}{i} = \log(n) + \gamma + o(1)$$

for $n \to \infty$.

Plesken and Robertz [82] generalised these results to A_n and to wreath products of groups with imprimitive action.

2.2.5 Generating Functions

One very powerful method of obtaining information about certain combinatorial quantities is to employ generating functions.

Given a sequence $(a_n)_{n \in \mathbb{N}}$ of real numbers, the *Ordinary Generating Function* for a_n is

$$A(z) := \sum_{n \geq 0} a_n z^n.$$

For example, a_n could be the number of certain elements in S_n.

A very intuitive way to view generating functions is given in the following quote from Wilf's aptly named book *generatingfunctionology* [96]: "A generating function is a clothesline on which we hang up a sequence of numbers for display." Here we just highlight some of the ways in which generating functions can shed light on some of our problems. To understand the power and beauty of the subject of generating functions we refer the reader to both Wilf's book [96] and a recent book on analytic combinatorics by Flajolet and Sedgewick [35]. Both books also contain various interesting results on proportions of permutations.

Several types of generating functions can be defined, and the type of generating function chosen to attack a particular problem depends on the circumstances. In our situation *exponential generating functions* are of particular interest. They are of the form

$$A(z) := \sum_{n \geq 0} \frac{a_n}{n!} z^n$$

and ensure that the coefficients $\frac{a_n}{n!}$ of z^n are manageable in situations where a_n is expected to grow almost as fast as $n!$. For example, if a_n is the number of elements with a particular property in S_n, then this number could grow rapidly and using an ordinary generating function would quickly produce unwieldy coefficients. However, dividing by the order of the group S_n ensures that the coefficients $a_n/n!$ are proportions of elements in S_n and thus all less than 1.

We study generating functions as elements of the ring of formal power series. Analytic questions, convergence etc. do not concern us just yet. Generating functions can be manipulated in various ways, and this theory is described in the above mentioned books. Here we just state, as an example, how two generating functions can be multiplied:

$$\left(\sum_{n=0}^{\infty} a_n z^n \right) \cdot \left(\sum_{n=0}^{\infty} b_n z^n \right) = \sum_{n=0}^{\infty} \left(\sum_{k=0}^{n} a_k b_{n-k} \right) z^n.$$

The usefulness of taking a generating function approach in our situation can be highlighted with the following example. A further example, that estimates the proportion of regular semisimple elements in general linear groups, is given in Sect. 2.3.6.

2.2.5.1 Example

Let $b \geq 1$ be a fixed integer and let a_n denote the number of permutations in S_n all of whose cycles have length at most b.

We would like to study the exponential generating function describing the numbers a_n. So let

$$A(z) := \sum_{n \geq 0} \frac{a_n}{n!} z^n.$$

One very effective way of studying a generating function is to start from a recursive equation for the coefficients a_n, and we employ this method here. Our first task is to find a suitable recursion for a_n. Recall that we write permutations in disjoint cycle notation. We are interested in finding an expression for the number a_n of permutations in S_n all of whose cycles have length at most b in terms of a_m for integers m smaller than n. We employ a combinatorial trick which has been used e.g. in Beals et al. [9, Theorem 3.7].

We enumerate the permutations in S_n all of whose cycles have length at most b according to the length d of the cycle containing the point 1. For a fixed d, we have $\binom{n-1}{d-1}$ choices for the remaining points of the cycle of length d containing 1. On these d points we can put any one of $(d-1)!$ different cycles and we have a_{n-d} choices for the permutation on the remaining $n-d$ points. Thus we obtain the recursion

$$\frac{a_n}{n!} = \frac{1}{n} \sum_{d=1}^{\min\{b,n\}} \frac{a_{n-d}}{(n-d)!}.$$

Note in particular that $a_n = n!$ for $n \leq b$, which is in agreement with this recursion. The recursion implies that

$$A(z) := \sum_{n=0}^{\infty} \frac{a_n}{n!} z^n = 1 + \sum_{n=1}^{\infty} \frac{1}{n} \left(\sum_{d=1}^{\min\{b,n\}} \frac{a_{n-d}}{(n-d)!} \right) z^n$$

$$= 1 + \sum_{d=1}^{b} \sum_{n=d}^{\infty} \frac{1}{n} \frac{a_{n-d}}{(n-d)!} z^n = 1 + \sum_{d=1}^{b} \sum_{n=0}^{\infty} \frac{1}{n+d} \frac{a_n}{n!} z^{n+d}.$$

A very useful trick when working with generating functions is to take the derivative. This yields in our case

$$A'(z) = \sum_{d=1}^{b} \sum_{n=0}^{\infty} \frac{a_n}{n!} z^{n+d-1} = \sum_{d=1}^{b} z^{d-1} \sum_{n=0}^{\infty} \frac{a_n}{n!} z^n = \sum_{d=1}^{b} z^{d-1} A(z).$$

Thus

$$\frac{A'(z)}{A(z)} = \sum_{d=1}^{b} z^{d-1}$$

and so

$$\log(A(z)) = \sum_{d=1}^{b} \frac{z^d}{d}.$$

Therefore we see that our generating function is

$$A(z) = \exp(\sum_{d=1}^{b} \frac{z^d}{d}).$$

While this has yielded a very succinct way of describing the number of elements of interest, it does not as yet yield the desired upper and lower bounds for the proportion of such elements. Thus we would like to know whether generating functions can tell us about the limiting behaviour of the coefficients.

An elaborate theory of the asymptotic behaviour of the coefficients of the generating functions exists. We mention here briefly a subject called "Saddlepoint Analysis". The theory is described in the above mentioned books (see also the papers by Moser and Wyman [67, 68] and Bender [11]). We quote here one result from Flajolet and Sedgewick's book, which helps in the situation of our example. The quoted result is based on a more general theorem by W.K. Hayman [43] (see also Theorem VIII.4 of [35]). In line with the literature, we denote the coefficient of z^n in the generating function $A(z)$ by $[z^n]A(z)$. The operator $z\frac{d}{dz}$ is defined by $z\frac{d}{dz} : P(z) \mapsto zP'(z)$.

Theorem 2.7 (see Corollary VIII.2 of [35]). *Let $P(z) = \sum_{j=1}^{n} a_j z^j$ have non-negative coefficients and suppose $\gcd(\{j \mid a_j \neq 0\}) = 1$. Let $F(z) = \exp(P(z))$. Then*

$$[z^n]F(z) \sim \frac{1}{\sqrt{2\pi\lambda}} \frac{\exp(P(r))}{r^n},$$

where r is defined by $rP'(r) = n$ and $\lambda = \left(z\frac{d}{dz}\right)^2 P(r)$.

2.2.5.2 Example of Saddlepoint Analysis

Recall that $A(z) = \exp(\sum_{d=1}^{b} \frac{z^d}{d})$ is the exponential generating function for the number of elements all of whose cycles have length at most b.

Let $P(z) = \sum_{d=1}^{b} \frac{z^d}{d}$. Then $P(z)$ is a polynomial in z with non-negative coefficients and satisfies $\gcd(\{d \mid \text{coefficient of } z^d \text{ is nonzero}\}) = 1$. The first step in applying Saddlepoint Analysis is to estimate r determined by the equation $n = rP'(r)$. We have $n = rP'(r) = r\sum_{d=1}^{b} r^{d-1} = \sum_{d=1}^{b} r^d \geq r^b$, and so $r \leq \sqrt[b]{n}$.

The next step is to estimate λ, where $\lambda = \left(r\frac{r}{dr}\right)^2 P(r) = r\sum_{d=1}^{b} dr^{d-1} = \sum_{d=1}^{b} dr^d \leq b\sum_{d=1}^{b} r^d = bn$.

Hence we have $r \leq n^{1/b}$, $\lambda \leq bn$ and $P(r) = \sum_{d=1}^{b} \frac{r^d}{d} \geq \frac{1}{b}\sum_{d=1}^{b} r^d = \frac{n}{b}$, implying

$$[z^n]A(z) \sim \frac{1}{\sqrt{2\pi\lambda}} \frac{\exp(P(r))}{r^n} \geq \frac{1}{\sqrt{2\pi bn}} \left(\frac{e}{n}\right)^{n/b}.$$

2.2.6   Solutions to $x^m = 1$ in Symmetric and Alternating Groups

The number of solutions to an equation of the form $x^m = 1$ for a fixed integer m in symmetric and alternating groups of degree n has received quite a lot of attention in the literature. More recently, interest in such equations has been rekindled due to algorithmic applications. In particular, it has also been important for algorithmic applications to find the asymptotic behaviour of the number of solutions of equations of the form $x^m = 1$ where m is allowed to grow with n.

We begin by outlining some of the results in the literature. For m fixed let

$$c(n, m) = \frac{1}{n!} |\{g \in S_n \mid g^m = 1\}|.$$

Let

$$C_m(z) = \sum_{n=0}^{\infty} c(n, m) z^n$$

be the corresponding generating function.

Theorem 2.8 (Jacobsthal [47]). *For a prime p we have*

$$C_p(z) = \exp\left(z + \frac{z^p}{p}\right) \quad and \quad c(n, m) = \sum_{\lambda=1}^{[n/p]} \frac{1}{(n - \lambda p)! \lambda! p^\lambda}.$$

The number $n! c(n, 2)$ of solutions to the equation $x^2 = 1$ in symmetric groups of degree n deserves particular attention, since it is also the sum of the degrees of the irreducible representations of S_n. Chowla et al. [20] examined $c(n, 2)$ and showed that $n \cdot c(n, 2) = c(n-1, 2) + c(n-2, 2)$. Thus they deduced that $1/\sqrt{n} \leq c(n, 2) \leq 1/\sqrt{n} + \frac{1}{n}$ and found the dominant term of the asymptotic expansion for $c(n, 2)$.

Later, Chowla et al. [21] were able to generalise Jacobsthal's expansion of $C_p(x)$ to $C_m(x)$ where m can be an arbitrary fixed integer, and they asked for an asymptotic formula for $c(n, m)$.

Theorem 2.9 (Chowla et al. [21]).

$$C_m(z) = \exp\left(\sum_{d \mid m} \frac{z^d}{d}\right).$$

Moser and Wyman [66, 67] derived an asymptotic formula in terms of a contour integral for $c(n, 2)$ and derived the first order term of the asymptotic value of $c(n, p)$. Moreover, they were able to obtain corresponding results for alternating groups.

Theorem 2.10 (Moser and Wyman [66,67]). *For a prime $p > 2$,*

$$c(n, p) \sim \frac{1}{n!} \frac{1}{\sqrt{p}} \left(\frac{n}{e}\right)^{n(1-\frac{1}{p})} e^{n\frac{1}{p}}.$$

Herrera [44] gives the following recursive formula for the number $n!b(n, m)$ of elements in S_n of order m:

$$n!b(n, m) = \sum_s \frac{(n-1)!}{(n-s)!} \sum_t b(n-s, t), \quad \text{where } \gcd(t, s) = m.$$

Other authors (e.g. Chernoff [19], Mineev and Pavlov [63], and Pavlov [80]) studied the number of elements in S_n or A_n satisfying an equation of the form $x^m = a$ for some element $a \in S_n$.

In 1986 Volynets [92], Wilf [95] and Pavlov [81] independently determined the limiting behaviour of $c(n, m)$ for fixed m, and n tending to infinity. The following theorem is Wilf's formulation of the result.

Theorem 2.11. *Let m be a fixed positive integer. Define $\varepsilon(n, m) = 0$ if m is odd and $\varepsilon(n, m) = 1/(2m^2n)$ if m is even; and let*

$$\tau = \frac{1}{n^{1/m}} \left(1 + \frac{1}{nm} \sum_{d | m, d < m} n^{d/n} + \varepsilon(n, m)\right).$$

Then for $n \to \infty$ we have

$$c(n, m) \sim \frac{\tau^n}{\sqrt{2\pi mn}} \exp\{\sum_{d | m} \frac{1}{d\tau^d}\}.$$

The above result has been generalised in the literature in various directions and we shall mention some of these.

2.2.6.1 Families of m

Ben-Ezra [10] generalised these formulae as follows. Let Π be a set of primes and let Π' denote the set of all primes not in Π. Further, let $C_\Pi(z)$ denote the generating function for the proportion $c(n, \Pi)$ of all elements whose order only involves primes in Π, and let $C_{\Pi'}(z)$ denote the generating function for the proportion $c(n, \Pi')$ of all elements whose order only involves primes in Π'. For a finite set B of integers, define $||B|| = 1$ if $B = \varnothing$ and $||B|| = \prod_{b \in B} b$ otherwise. Then

Theorem 2.12 (Ben-Ezra [10]).

1. $C_{\Pi}(z) = \prod_{\substack{B \subseteq \Pi' \\ |B| < \infty}} (1 - z^{||B||})^{\frac{(-1)^{|B|+1}}{||B||}}$.

2. $C_{\Pi'}(z) = \prod_{\substack{B \subseteq \Pi \\ |B| < \infty}} (1 - z^{||B||})^{\frac{(-1)^{|B|+1}}{||B||}}$.

2.2.6.2 Growing m

The first author to consider an equation of the form $x^m = 1$ in symmetric groups of degree n in which m is not assumed to be fixed was Warlimont [93], who considers the case $m = n$. In particular, he shows that

Theorem 2.13 (Warlimont [93]).

$$\frac{1}{n} + \frac{2c}{n^2} \le c(n, n) \le \frac{1}{n} + \frac{2c}{n^2} + O\left(\frac{1}{n^{3-o(1)}}\right),$$

where $c = 0$ if n is odd and $c = 1$ if n is even.

In 1990 Erdős and Szalay [26] considered the case where m lies in the range $\log(n)/(2 \log \log(n)) \le m \le n^{(1/4)-\varepsilon}$, and derived an asymptotic formula for $c(n, m)$.

Volynets [92] proved the following result via the Saddlepoint method.

Theorem 2.14 (Volynets [92]). *For primes p, and for positive integers n such that n and p tend to ∞ and $p/n \to 0$,*

$$c(n, p) = \frac{1}{n!} \left(\frac{n}{e}\right)^{n(1-1/p)} p^{1/2} \sum_{k=0}^{\infty} \frac{(n^{1/p})^{m+kp}}{(m + kp)!} (1 + o(1)),$$

where $m = n - p[n/p]$. In particular, if $n^{1/p}/p^2 \to 0$ then

$$c(n, p) = \frac{1}{n!} \left(\frac{n}{e}\right)^{n(1-1/p)} p^{1/2} e^{n^{1/p}} (1 + o(1)),$$

while if $n^{1/p}/p \to 0$ then

$$c(n, p) = \frac{1}{n!} \left(\frac{n}{e}\right)^{n(1-1/p)} p^{1/2} \frac{n^{m/p}}{m!} (1 + o(1)).$$

Finally A.V. Kolchin [49] proved the following theorem using the method of generalised schemes of allocation (see [50, Chap. 5]).

Theorem 2.15 (Kolchin [49]). *For d, n positive integers such that $d \log\log(n)/\log(n) \to 0$ and for $\delta = 0$ if d is odd and $\delta = 1/(2d)$ when d is even, the following holds:*

$$c(n,d) = \frac{1}{n!} \frac{n^{n(1-1/d)}}{e^n} \frac{1}{\sqrt{d}} \exp\left\{ \sum_{j|d} \frac{n^{j/d}}{j} - \delta \right\} (1 + o(1)).$$

Another generalisation of the above to the case where the cycle lengths are elements of particular sets can be found in [90]. Finally we would like to refer the interested reader to V.F. Kolchin's book on random graphs [50], which contains many references and notes to the above mentioned, and other, results on random permutations.

2.2.7 The Münchausen Method (Bootstrapping)

The previous results highlight how difficult it is to obtain the overall limiting behaviour for $c(n,m)$ when $m \leq \ell n$ for some constant ℓ and m is allowed to grow with n. However, for our algorithmic applications (see Sect. 2.2.8 below), we require good upper bounds for $c(n,m)$ in the case where $m = r(n-k)$ for $r \in \{1,2,3\}$ and $k \leq 6$.

To obtain bounds for $c(n,m)$ in cases where $n-1 \leq m \leq \ell n$ for some constant ℓ, we return to more basic methods and highlight some of the ideas in a proof of the limiting behaviour of $c(n,m)$ in such cases.

A popular folk tale tells the story of how Baron Münchausen found himself stuck in a swamp while riding his horse. He then managed to save himself and his horse by pulling himself out of the swamp by his own ponytail.

We employ a similar strategy to obtain good estimates for our required proportions. We begin by deriving a first crude estimate and then using this to refine our estimates. This method (also called bootstrapping) was employed in [9] and later in [73].

The overall estimate for $c(n,m)$ is obtained in two steps. The first step yields a very crude estimate. This in turn is employed in a second step to yield a more refined estimate.

$$\text{Define } \gamma(m) := \begin{cases} 2 & \text{for } 360 < m \\ 2.5 & \text{for } 60 < m \leq 360 \\ 3.345 & \text{for } m \leq 60. \end{cases}$$

A first crude estimate for $c(n,m)$ is given in the following theorem.

Theorem 2.16. *Let $m, n \in \mathbb{N}$ with $m \geq n - 1$. Then*

$$c(n,m) \leq \frac{1}{n} + \frac{\gamma(m)m}{n^2}.$$

Proof-Idea for Crude Estimate

The proof of our first crude estimate relies on a simple idea. It divides the problem of estimating $c(n, m)$ into several smaller problems by considering the following proportions in S_n (see [9]) according to how many cycles the numbers 1, 2 and 3 lie in. Define proportions

1. $c^{(1)}(n, m)$ of those $g \in S_n$ which have 1, 2, 3 in the same g-cycle.
2. $c^{(2)}(n, m)$ of those $g \in S_n$ which have 1, 2, 3 in two g-cycles.
3. $c^{(3)}(n, m)$ of those $g \in S_n$ which have 1, 2, 3 in three g-cycles.

Then it is clear that

$$c(n, m) = c^{(1)}(n, m) + c^{(2)}(n, m) + c^{(3)}(n, m).$$

For each i with $i \in \{1, 2, 3\}$, we can now hope to use the extra knowledge about the elements that contribute to the proportion $c^{(i)}(n, m)$ to obtain a first estimate for this proportion.

For example, we show how we can obtain an estimate for $c^{(1)}(n, m)$. Elements $g \in S_n$ contributing to this proportion must contain a cycle C of length d with the following properties:

1. $d \mid m$ and $3 \le d$.
2. The cycle C of length d contains 1,2,3.
3. The remaining cycles of g all have lengths dividing m.

Now we can obtain an expression for $c^{(1)}(n, m)$ by considering all allowable cycle lengths d and counting the number of cycles C on d points that contain the points 1, 2 and 3 and ensuring that the remaining $n - d$ points all have lengths dividing m. As C has to contain 1, 2 and 3, we have $n - 3$ points left to choose the remaining $d - 3$ points of C; and having chosen a set of d points (which contains the points 1, 2 and 3), we have $(d - 1)!$ ways of arranging them into different cycles. The number of permutations on the remaining $n - d$ points all of whose cycle lengths divide m is $c(n - d, m)(n - d)!$. Hence

$$c^{(1)}(n, m) = \frac{1}{n!} \sum_{d \mid m, d \ge 3} \binom{n-3}{d-3} (d-1)! c(n-d, m)(n-d)!$$

$$= \frac{(n-3)!}{n!} \sum_{d \mid m, 3 \le d \le n} (d-1)(d-2) c(n-d, m).$$

As we are currently only interested in obtaining a first crude estimate, we apply a very rough upper bound on $c(n - d, m)$, by replacing it with the constant 1. We therefore find

$$c^{(1)}(n,m) \leq \frac{(n-3)!}{n!} \sum_{d\mid m, 3 \leq d \leq n} (d-1)(d-2)$$

$$\leq \frac{(n-3)!}{n!} \sum_{m/n \leq t \leq m/3} \left(\frac{m}{t}-1\right)\left(\frac{m}{t}-2\right)$$

$$\leq \frac{(n-3)!}{n!} \left((n-1)(n-2) + \int_{m/n}^{m/3} \frac{m^2}{t^2}\, dt \right)$$

$$\leq \frac{(n-3)!}{n!} \{(n-1)(n-2) + mn - 3m\}$$

$$< \frac{1}{n} + \frac{m}{n^2}.$$

We can employ similar estimates to obtain crude upper bounds for $c^{(2)}(n,m)$ and $c^{(3)}(n,m)$, which we omit here. Having obtained a first crude estimate, we now insert this estimate when trying to get a better estimate for $c(n,m)$.

2.2.7.1 The Pull

Enumerating g by the g-cycle of length d on 1 and recalling that $n-1 \leq m$ yields

$$c(n,m) = \frac{1}{n} \sum_{\substack{d\mid m \\ 1 \leq d \leq n}} c(n-d,m)$$

$$\leq \frac{1}{m} + \frac{1}{n} \sum_{\substack{d\mid m \\ 1 \leq d \leq m/2}} c(n-d,m).$$

For example, in the case where $m = n$ or $m = n-1$, inserting the crude estimate for $c(n-d,m)$ in the equations above we find that

$$c(n,m) \leq \frac{1}{m} + \frac{1}{n} \sum_{\substack{d\mid m \\ 1 \leq d \leq m/2}} \left(\frac{1}{n-d} + \frac{\gamma(m)m}{(n-d)^2} \right)$$

$$\leq \frac{1}{m} + \frac{d(m)(2 + 4\gamma(m))}{n^2},$$

where $d(m)$ denotes the number of positive integer divisors of m. The above results allow us to prove the following strong corollaries.

Corollary 2.17. *Let $n \geq 19$. Let $f \in \{n-3, n-2\}$ be odd. Then*

1. *The conditional probability that a random element g has an n-cycle given that it satisfies $g^n = 1$ is at least $1/2$.*

2. *The conditional probability that a random element g has an f-cycle given that it satisfies $g^{2f} = 1$ and $|g^f| = 2$ is at least $1/3$.*

Finally, we highlight some of the results proved in [75] estimating $c(n,m)$, where $m = rn$ for a fixed value of r. The proof of this theorem relies on ideas similar to those outlined above, combined with an idea of Warlimont's [93] dividing cycles of permutations into large and small cycles.

Theorem 2.18. *For positive integers r, n with r fixed and n sufficiently large,*

$$c(n, rn) = \frac{1}{n} + \frac{a(r)}{n^2} + O\left(\frac{1}{n^{\frac{5}{2}-o(1)}}\right)$$

where $a(r) = \sum_{i,j}(1 + \frac{i+j}{2r})$, $1 \le i, j \le r^2$, $ij = r^2$ and $r + i, r + j$ divide rn. Moreover, the conditional probability that an element $g \in S_n$ is an n-cycle, given that its order divides rn, is at least $1 - \frac{a(r)}{n} - O\left(\frac{1}{n^{3/2-o(1)}}\right)$.

2.2.8 Algorithmic Applications of Proportions in Symmetric Groups

Warlimont's result is very useful for algorithmic purposes. It tells us that most permutations g satisfying the equation $g^n = 1$ are n-cycles. Moreover, it also identifies the cycle structure of the second most abundant set of permutations satisfying the equation $g^n = 1$; namely permutations which consist of two cycles of length $n/2$, and these only occur when n is even. This translates into the algorithm below to find an n-cycle. Note that the algorithm works in *any permutation or matrix group representation* of S_n, where we may not easily recognise the cycle structure of an element in the natural representation. Such algorithms are called *black box group algorithms*; for a formal definition, see Sect. 2.4.2.

Suppose we are given a group G and we believe G might be isomorphic to S_n under a putative, yet unknown, isomorphism $\lambda : G \to S_n$. We find an element $g \in G$ which would map to an n-cycle under λ with high probability by Algorithm 3 below.

Algorithm 3: FINDNCYCLE

    Input: G a group, $n \ge 19$ an integer, $0 < \varepsilon < 1$ real;
    Output: **g** or **fail**;
    # If the output is **g**, then $g^n = 1$;
    for *up to $n \log(\varepsilon^{-1})$ random elements $g \in G$* **do**
        if $g^n = 1$ **then**
            return g;
        end
    end
    Return **fail**;

The algorithm takes as input a real ε such that $0 < \varepsilon < 1$, and this input is used to control the probability of failure. We require that the probability that G is isomorphic to S_n and the algorithm returns **fail** to be at most ε. Note that on each random selection, the probability of finding an n-cycle is $1/n$. Hence the probability of failing to find an n-cycle in $N(\varepsilon)$ random selections is $(1 - 1/n)^{N(\varepsilon)}$ and we have $(1 - 1/n)^{N(\varepsilon)} < \varepsilon$ when $N(\varepsilon) \geq \log(\varepsilon^{-1})/(-\log(1 - 1/n))$. In particular, this is the case when $N(\varepsilon) \geq n \log(\varepsilon^{-1})$.

Thus the above algorithm returns with probability at least $1 - \varepsilon$ an element $g \in G$ satisfying $g^n = 1$. Therefore, if $G \cong S_n$ then with probability at least $1/2$ this element is an n-cycle, by the above corollary.

Niemeyer and Praeger [74] generalise Warlimont's result and consider the case where $m \geq n$, namely $rn \leq m < (r + 1)n$ for fixed positive integers r.

Algorithm 3 is part of a procedure which decides whether a black box group G is isomorphic to the full symmetric group S_n for a given natural number n. The full algorithm is described in [9]. First, we have to describe a presentation for the group S_n.

Theorem 2.19 (Coxeter and Moser, 1957).

$$\langle r, s \mid r^n = s^2 = (rs)^{(n-1)} = [s, r^j]^2 = 1 \ for \ 2 \leq j \leq n/2 \rangle$$

is a presentation for S_n. Moreover, if some group G has generators r, s satisfying this presentation and $r^2 \neq 1$ then G is isomorphic to S_n.

Definition 2.20. The transposition y *matches* the n-cycle x if y moves two adjacent points in x.

Lemma 2.21. *For $n \geq 5$, an n-cycle and a matching transposition satisfy the presentation in Theorem 2.19.*

Now we are ready to sketch the algorithm BBRECOGNISESN of [9].

Algorithm 4: BBRECOGNISESN

Input: $G = \langle X \rangle$ a black box group, $n \geq 5$;
Output: true and a map $\lambda : G \to S_n$, **or fail**;
repeat
    1. find $r \in G$ with $r^n = 1$.
       # is $\lambda(r)$ an n-cycle?
    2. find $h \in G$ with $h^{2m} = 1$ where $m \in \{n - 2, n - 3\}$ odd.
       # is $\lambda(h^m)$ a transposition?
    3. find a random conjugate s of h^m with $[s, s^g] \neq 1$.
       # does $\lambda(s)$ interchange two points of $\lambda(r)$?

until *repeated too often*;
if *r or s not found* **then return fail**;
else
    define λ by

    • $\lambda(r) = (1, \ldots, n)$ and
    • $\lambda(s) = (1, 2)$.

    Return true and $\lambda : G \to S_n$;
end

We test whether $\langle r, s \rangle \cong S_n$ via the presentation described in Theorem 2.19.

Theorem 2.22. *Given a black box group G isomorphic to S_n, the probability that the algorithm* BBRECOGNISESN(G, n, ε) *returns* **fail** *is at most ε. The cost of the algorithm is*

$$O((n\xi + n \log(n)\mu) \log(\varepsilon^{-1})),$$

where ξ is the cost of finding a random element in a black box group and μ the cost of a black box group operation.

2.2.9 Restrictions on Cycle Lengths

An extensive amount of literature exists on the topic of random permutations whose cycle lengths lie in a given set $\mathscr{L}$ or lie in a particular arithmetic progression. Early work includes that of Touchard [91], Gončarov [40] and Gruder [42].

Let $\mathscr{L}$ be a set of natural numbers. Let $d_{\mathscr{L}}(n)$ denote the proportion of elements in S_n all of whose cycle lengths lie in $\mathscr{L}$ and let $d_{\mathscr{L}}(n, k)$ denote the proportion of elements in S_n with exactly k cycles all of whose lengths lie in $\mathscr{L}$. A generating function for $d_{\mathscr{L}}(n)$ can be found in [91]. This proportion has been studied by many authors; we just mention briefly some of Gruder's results.

Theorem 2.23 (Gruder [42]).

$$d_{\mathscr{L}}(n, k) = \frac{1}{k!} \sum_{\substack{(x_1, \dots, x_k) \in \mathscr{L}^k \\ x_1 + \dots + x_k = n}} \frac{1}{x_1 \cdots x_k}.$$

Put $H(z) = \sum_{a \in \mathscr{L}} \frac{z^a}{a}$ and let $D(z) = \sum_{n=0}^{\infty} d_{\mathscr{L}}(n) z^n$.

Theorem 2.24 (Gruder [42]).

1. $D(z) = \exp(H(z))$.
2. $D(z)^x = \exp(x H(z)) = \sum_{n=0}^{\infty} \left(\sum_{k=0}^{n} d_{\mathscr{L}}(n, k) x^k \right) z^n$.

Bolker and Gleason [13] obtain an explicit asymptotic formula for $d_{\mathscr{L}}(n)$ when $\mathscr{L}$ is an arithmetic progression.

Let $p_a(n)$ denote the proportion of elements in S_n all of whose cycle lengths are at least a for some $a \geq 2$.

Theorem 2.25 (Gruder [42]).

1. $\lim_{n \to \infty} \frac{1}{p_a(n)} = \exp(1 + \frac{1}{2} + \dots + \frac{1}{a-1})$.
2. $\log \left(\lim_{a \to \infty} \lim_{n \to \infty} \frac{1}{p_a(n)} \right) = \gamma$, *where $\gamma = \lim_{n \to \infty} \left(\sum_{i=1}^{n} \frac{1}{i} - \log(n) \right)$ is the Euler constant.*

V.F. Kolchin summarises many of the asymptotic results known about this case in his book [50]. We refer the interested reader to [50] and references therein.

Finally, we mention one particular proportion that has been of considerable interest in various applications. For positive integers b, let $p_{\neg b}(n)$ denote the proportion of elements in S_n with no cycle of length divisible by b. This proportion was first studied for primes b in [28], where Erdős and Turán give an explicit formula for it. This formula immediately generalises to arbitrary positive integers b. For a prime b, Erdős and Turán also give the limiting distribution of $p_{\neg b}(n)$. Many other authors have also considered this proportion; for example [12], [14, Sect. 2], [38]. Here we quote a result from [8, Theorem 2.3(b)].

Theorem 2.26. *Let* $n \geq b$. *Then*

$$
\left(\frac{b}{n}\right)^{1/b} \frac{(1 - \frac{1}{n})}{\Gamma(1 - \frac{1}{b})} \leq p_{\neg b}(n) \leq \left(\frac{b}{n}\right)^{1/b} \frac{(1 + \frac{2}{n})}{\Gamma(1 - \frac{1}{b})}.
$$

Ben-Ezra [10] obtained a similar result for $b = 2$. A formula for the proportion of elements in A_n with no cycle of length divisible by b is also given in [8]. Maróti [61] generalises this, and gives a formula for the proportion of elements of order not divisible by b in arbitrary permutation groups.

The above estimates have proved to be very useful in deriving proportions of certain elements in finite classical groups of Lie type. Suppose G is a finite classical group of Lie type given in natural dimension n with $n \geq 2$. Using the method outlined in Sect. 2.3.4, [58] shows that the proportion of elements in G for which some power is an involution with a large 1-eigenspace of dimension d with $n/3 \leq d \leq 2n/3$ is at least $c/\log(n)$ for some constant c.

2.3 Estimation Techniques in Lie Type Groups

We start with a seemingly simple result about permutation groups, discuss the deep Lie-theoretic analysis underpinning it, and indicate how this approach has led to a powerful estimation technique for Lie type groups.

2.3.1 p-Singular Elements in Permutation Groups

The following beautiful and surprising result of Isaacs et al. [46] was published in 1995.

Theorem 2.27 (Isaacs, Kantor and Spaltenstein [46]). *Let* $G \leq S_n$ *and let* p *be a prime dividing* $|G|$. *Then there is at least* 1 *chance in* n *that a uniformly distributed random permutation in* G *has order a multiple of* p.

This result is about *any permutation group*—not necessarily primitive, nor even transitive. It is best possible for such a general result, since if $n = p$ then in the affine group $\mathrm{AGL}(1, p)$ there are exactly $p - 1$ elements of order divisible by p out of a total of $p(p - 1)$ elements in the group.

The only known proof of Theorem 2.27 requires the finite simple group classification. The proof strategy is first to make an elementary reduction to the case where G is a nonabelian simple group. Then the simple groups are dealt with. There are no difficulties with the alternating groups A_n and the sporadic simple groups. This leaves the finite simple groups of Lie type to be considered, and this is where the authors of [46] "wave a magic wand" with a sophisticated argument from the theory of Lie type groups. We (Niemeyer and Praeger) were at first baffled by this proof, as well as fascinated by what it achieved, so set about trying to understand it. Along the way there was help from Klaus Lux and Frank Lübeck. With Frank Lübeck we made our first full-blown application of the theory in [58] to estimate the proportion of a certain family of even ordered elements in classical groups. We discovered that this beautiful theory had been introduced by Gus Lehrer [54,55] to count various element classes and representation theoretic objects associated with Lie type groups. Recently Arjeh Cohen and Scott Murray [22] also used this approach to develop algorithms for computing with finite Lie algebras.

Our objective became: to formalise the ideas into a framework for estimating proportions of a wide class of subsets of finite Lie type groups. The framework was first set out in [58] and in general in [76]. We describe it in the next subsection.

2.3.2   Quokka Subsets of Finite Groups

For a finite group G and a prime p dividing $|G|$, each group element g can be written uniquely as a commuting product $g = us = su$, where u is a p-element and s is a p'-element (that is, $\mathrm{ord}(u)$ is a power of p while $\mathrm{ord}(s)$ is coprime to p). This is called the *Jordan p-decomposition* of g.

To find this decomposition write $\mathrm{ord}(g) = p^a b$ where $p \nmid b$ and $a \geq 0$. Then since p^a and b are coprime, there are integers r, t such that $rp^a + tb = 1$. It is straightforward to check that the elements $u = g^{tb}$ and $s = g^{rp^a}$ have the required properties, and that u, s are independent of the choices for r, t. This decomposition is critical for defining the kinds of subsets amenable to this approach for estimation.

Definition 2.28. Let G be a finite group and p a prime dividing $|G|$. A non-empty subset Q of G is called a *quokka set*, or a *p-quokka set* if we wish to emphasise the prime p, if the following two properties hold:

(a) Q is closed under conjugation by elements of G.
(b) For $g \in G$ with Jordan p-decomposition $g = us = su$, $g \in Q$ if and only if $s \in Q$.

A natural place to find p-quokka sets is in finite Lie type groups in characteristic p; for example, in $G = \mathrm{GL}(n, q)$ with q a power of p. Here, in a Jordan p-decomposition $g = us = su$, the element u is unipotent and s is semisimple. The elements u, s are called the *unipotent part* and the *semisimple part* of g, respectively. Some of the subsets already discussed in this chapter turn out to be quokka sets. We give an example.

Example 2.29. Let $G = \mathrm{GL}(n, q)$ or $\mathrm{SL}(n, q)$, with q a power of p, let e be an integer such that $e > n/2$, and suppose that $q^e - 1$ has a primitive prime divisor. Then the subset Q of ppd-$(n, q; e)$ elements of G is a p-quokka set. To see this, note that Q is closed under conjugation since conjugate elements have the same order. Also, for a Jordan p-decomposition $g = us = su$, a ppd r of $q^e - 1$ divides $\mathrm{ord}(g)$ if and only if r divides $\mathrm{ord}(s)$.

2.3.3 Estimation Theory for Quokka Sets

The standard reference for the concepts discussed below is Roger Carter's book [17], and an account of the required theory is given in [76].

The groups: We start with a connected reductive algebraic group G defined over the algebraic closure $\overline{\mathbb{F}_q}$ of the finite field $\mathbb{F}_q$ of order q, where q is a power of a prime q_0. A Frobenius morphism $F : G \to G$ defines a finite group of Lie type $G^F = \{g \in G \mid F(g) = g\}$ as its fixed point subgroup. We use the following example to illustrate the concepts as they arise. For the algebraic group $G = \mathrm{SL}(n, \overline{\mathbb{F}_q})$ and Frobenius morphism $F : (a_{ij}) \mapsto (a_{ij}^q)$, the finite group of Lie type is $G^F = \mathrm{SL}(n, q)$, since the fixed field of the map $a \mapsto a^q$ is $\mathbb{F}_q$.

Maximal tori: A *torus* in an algebraic group is a subgroup T that is isomorphic to a direct product of a finite number of copies of the multiplicative group of $\overline{\mathbb{F}_q}$. In particular, T is abelian. A torus T is F-stable if $F(T) = T$, and T is a maximal torus if T is closed and not properly contained in another torus. All F-stable maximal tori in G are conjugate. In our example $G = \mathrm{SL}(n, \overline{\mathbb{F}_q})$, the subgroup T_0 of diagonal matrices in G is a maximal torus that is isomorphic to a direct product of $n - 1$ copies of $(\overline{\mathbb{F}_q})^*$.

The Weyl group: Choose an F-stable maximal torus T_0 in G. The *Weyl group* W is defined as the quotient $N_G(T_0)/T_0$. Since F-stable maximal tori are conjugate, the group W is independent of the choice of T_0. In our example $G = \mathrm{SL}(n, \overline{\mathbb{F}_q})$, with T_0 the subgroup of diagonal matrices, $N_G(T_0)$ is the subgroup of monomial matrices in G, and $W = N_G(T_0)/T_0$ is isomorphic to the group of $n \times n$ permutation matrices, so $W \cong S_n$.

F-conjugacy: Elements $v, w \in W$ are said to be *F-conjugate* if there is an element $x \in W$ such that $v = x^{-1} w F(x)$. Notice that we abuse notation a little in this definition, since $x \in W$ is a coset $x = x_0 T_0$ and by $F(x)$ we mean $F(x_0) T_0$

(which is well defined since T_0 is F-stable). In our example $G = \mathrm{SL}(n, \overline{\mathbb{F}_q})$, F-conjugation is ordinary conjugation (since each $x \in W$ has a representative monomial matrix with entries 0 or ± 1, and hence x is fixed by F).

A crucial correspondence and the Quokka Theorem: For an F-stable maximal torus T of G, the intersection $T^F = T \cap G^F = \{g \in T \mid F(g) = g\}$ is called a *maximal torus* of G^F; although all F-stable maximal tori of G are G-conjugate, there are usually several G^F-conjugacy classes of F-stable maximal tori T^F, and the structure of the T^F is governed by the Weyl group. *There is a 1–1 correspondence between G^F-conjugacy classes of F-stable maximal tori and F-conjugacy classes of the Weyl group.* This is a crucial ingredient in proving the main theorem below. Let $\mathscr{C}$ be the set of F-conjugacy classes in W, and for $C \in \mathscr{C}$, let T_C^F denote a representative F-stable maximal torus of G^F corresponding to C.

Theorem 2.30. *Let G, F, T_0, W and $\mathscr{C}$ be as above, and let $Q \subset G^F$ be a quokka set. Then*

$$\frac{|Q|}{|G^F|} = \sum_{C \in \mathscr{C}} \frac{|C|}{|W|} \cdot \frac{|T_C^F \cap Q|}{|T_C^F|}.$$

Bounds on proportions: Essentially Theorem 2.30 allows us to separate an estimation problem within a Lie type group G^F into two simpler problems, one within the Weyl group and the other within various maximal tori. The expression for $\frac{|Q|}{|G^F|}$ in Theorem 2.30 as an exact summation can lead to usable bounds. Suppose that $\hat{\mathscr{C}}$ is a union of F-conjugacy classes and that ℓ_Q is a positive constant such that $\frac{|T_C^F \cap Q|}{|T_C^F|} \geq \ell_Q$ for all $C \in \hat{\mathscr{C}}$. Then Theorem 2.30 implies that $\frac{|Q|}{|G^F|} \geq \ell_Q \frac{|\hat{\mathscr{C}}|}{|W|}$. Similarly, if u_Q is such that $\frac{|T_C^F \cap Q|}{|T_C^F|} \leq u_Q$ for all $C \in \hat{\mathscr{C}}$, then $\frac{|Q|}{|G^F|} \leq u_Q \frac{|\hat{\mathscr{C}}|}{|W|}$.

A worked example: Let $G = \mathrm{SL}(n, \overline{\mathbb{F}_q})$ and let Q be the quokka set of ppd-$(n, q; e)$ elements of G, for some $e \in (n/2, n)$—see Example 2.29. We use this "quokka theory" to re-derive Lemma 2.3. The Weyl group is $W \cong S_n$, and each maximal torus T_C^F containing an element of Q is of the form

$$T_C^F = Z_{q^e-1} \times \text{other cyclic factors.} \tag{2.1}$$

As we discussed in the last paragraph of the proof of Lemma 2.3, for each such torus, the proportion $\frac{|T_C^F \cap Q|}{|T_C^F|}$ lies between $1 - \frac{1}{e+1}$ and 1. The F-conjugacy class C in W corresponding to such a torus consists of certain elements of $W = S_n$ containing an e-cycle, and all classes C with this property correspond to tori T_C^F as in (2.29). Let $\hat{\mathscr{C}}$ be the subset of W of all elements containing an e-cycle. Then $|\hat{\mathscr{C}}|/|W| = 1/e$, and as we discussed above, $|Q|/|G^F|$ lies between $(1 - \frac{1}{e+1})\frac{1}{e} = \frac{1}{e+1}$ and $\frac{1}{e}$.

r-abundant elements: The original impetus to study the work of Isaacs et al. [46] so closely came from efforts of Niemeyer and Praeger to understand whether, for a prime r, the lower bound given in [46] for the proportion of r-singular elements in finite classical groups was close to the true proportion. (An *r-singular element* is one with order a multiple of r.) Niemeyer conducted a computer experiment on general linear groups $G = \mathrm{GL}(n, p^a)$, for various dimensions n and primes p and r, where r divides $|G|$ and $r \neq p$, to discover the kinds of r-singular elements in G which appeared frequently on repeated independent random selections from G. It turned out that a good proportion of the r-singular elements that we found left invariant, and acted irreducibly on, a subspace of dimension greater than $n/2$. Moreover, their frequency seemed to be roughly proportional to $1/e$, where e is the smallest positive integer such that r divides $p^{ae} - 1$. We decided to call these elements *r-abundant*. It seemed at first that the r-abundant elements alone occurred with frequency greater than the lower bound predicted in [46]. However, it was pointed out to us by Klaus Lux that, hidden in the proofs in [46] was a lower bound on the proportion of r-singular elements of the form c/e for some constant c, with e as above. If $e > n/2$ then these r-singular elements are the ppd-$(n, p^a; e)$ elements used in the classical recognition algorithm in [72], and in general r-abundant elements are as easily recognisable as ppd elements from properties of their characteristic polynomials: namely, there is an irreducible factor $f(x)$ of degree greater than $n/2$ and a multiple of e, such that x has order a multiple of r modulo $f(x)$ in the polynomial ring $\mathbb{F}_{p^a}[x]$. A detailed study of r-abundant elements was carried out by Niemeyer and Praeger with Tomasz Popiel [71] to prove that the experimentally observed proportion of r-singular elements in general linear groups is correct, and to find and prove analogues for other finite classical groups. The r-abundant elements form a quokka set, and their proportion was determined [71, Theorem 1.1] using Theorem 2.30. For the general linear group $\mathrm{GL}(n, p^a)$, the proportion is

$$\left(1 - \frac{1}{r^{t-1}(r+1)}\right) \cdot \frac{\ln(2)}{e}$$

with an error term of the form c/n for some constant c, where r^t is the largest power of r dividing $p^{ae} - 1$. It would be interesting to know if r-abundant elements could be useful algorithmically to identify classical groups. To aid our understanding of such elements, Sabina Pannek is undertaking a Ph.D. project to find which maximal subgroups of finite classical groups contain elements with an irreducible invariant subspace of the natural module of more than half the dimension.

2.3.4 Strong Involutions in Classical Groups

In [53], Leedham-Green and O'Brien introduced a new Las Vegas algorithm to find standard generators for a finite simple n-dimensional classical group H in odd characteristic in its natural action. (Recall that a randomised algorithm is called *Las Vegas* if the output, if it exists, is always correct; the algorithm may report failure with a small probability.) The algorithm of [53] proceeds by constructing recursively various centralisers of involutions (elements of order 2), the details of

Table 2.1 The classical
groups for Theorem 2.31 and
Corollary 2.32

S	X	n
$SL(\ell+1,q)$	$GL(\ell+1,q)$	$\ell+1$
$SU(\ell+1,q)$	$GU(\ell+1,q)$	$\ell+1$
$Sp(2\ell,q)$	$GSp(2\ell,q)$	2ℓ
$SO(2\ell+1,q)$	$GO(2\ell+1,q)$	$2\ell+1$
$SO^{\pm}(2\ell,q)$	$GO^{\pm}(2\ell,q)^0$	2ℓ

which are discussed further in Sect. 2.4.3. The issue we address here is how to find
an appropriate involution. Leedham-Green and O'Brien wished to work with an
involution whose centraliser would be essentially a product of two smaller classical
groups, each of roughly half the dimension. They called such involutions "strong":
an involution is *strong* if its fixed point subspace has dimension in $[n/3, 2n/3)$,
or equivalently if its -1-eigenspace has dimension in $(n/3, 2n/3]$. Let I denote
the subset of strong involutions in H. Leedham-Green and O'Brien constructed
elements of I by making independent, uniformly distributed random selections from
H to find an element of even order which powered up to a strong involution. We call
such elements *preinvolutions*. To estimate how readily a preinvolution can be found
by random selection, we need to estimate the size of the set

$$P(H, I) = \{h \in H \mid \text{ord}(h) \text{ is even, } h^{\text{ord}(h)/2} \in I\}. \tag{2.2}$$

Leedham-Green and O'Brien estimated that it would require $O(n\xi + n^4 \log n + n^4 \log q)$ elementary field operations (that is, additions, multiplications or inver-
sions) to compute a strong involution in H, where ξ is an upper bound on
the number of elementary field operations required to produce an independent,
uniformly distributed random element of H; see [53, Theorem 8.27]. Underpinning
this complexity estimate was their estimate that the proportion of preinvolutions in
H was at least c/n, for a constant c.

Niemeyer and Praeger, with Frank Lübeck, used the approach described in
Sect. 2.3.3 to obtain an improved estimate for this proportion [58, Theorem 1.1].
They considered any n-dimensional classical group H satisfying $S \leq H \leq X$,
where S, X, n are as in one of the lines of Table 2.1 with q odd. Here
$GO^{\pm}(2\ell,q)^0$ denotes the connected general orthogonal group—the index 2 subgroup of $GO^{\pm}(2\ell,q)$ that does not interchange the two $SO^{\pm}(2\ell,q)$-classes of
maximal isotropic subspaces.

Theorem 2.31. *Let H satisfy $S \leq H \leq X$, with S, X, n as in one of the lines of
Table 2.1, with q odd and $\ell \geq 2$, and let $I \subset H$ be the set of strong involutions.
Then*

$$\frac{|P(H, I)|}{|H|} \geq \frac{1}{5000 \log_2(\ell)}.$$

The weak constant of $1/5000$ arises from the fact that the estimation only
considered one class of elements that power up to a strong involution, and from the
fact that it determined one constant that is valid uniformly for all classical groups.

A more detailed analysis taking into account a wider family of preinvolutions would yield a larger value for the constant.

Lübeck, Niemeyer and Praeger also obtained similar lower bounds for projective groups: note that, for $Z_0 \leq Z(X)$, since the subset I of involutions in Theorem 2.31 contains no central elements, the set $\overline{I} := I Z_0 / Z_0$ is a subset of involutions in the projective group $\overline{H} := H Z_0 / Z_0$.

Corollary 2.32. *With the above notation,* $|P(\overline{H}, \overline{I})| / |\overline{H}| \geq 1/(5000 \log_2 \ell)$.

Using this new lower bound reduces the complexity of computing a strong involution in [53] to $O(\log(n)\xi + n^4 \log n + n^4 \log q)$; that is, replacing the first summand $n\xi$ by $\log(n)\xi$. It seems to be typical that whenever "quokka theory" is applicable, it produces superior estimates to more intuitive geometric methods.

In Sect. 2.4.3, the algorithm in [53] will be discussed further. Here we just mention that the proof of [58, Theorem 1.1] could have been given for a more general class of involutions called "balanced involutions". For constants α, β such that $0 < \alpha < 1/2 < \beta < 1$, an (α, β)-*balanced involution* in an n-dimensional classical group H is one with fixed point subspace having dimension in $[\alpha n, \beta n)$. The resulting lower bound on the proportion of (α, β)-balanced involutions in H would be $c / \log_2(n)$, for a constant c depending only on α and β.

2.3.5 More Comments on Strong Involutions

Before leaving this topic we make some comments about the proof of Theorem 2.31. First, it is not difficult to see that $P(H, I)$ is a quokka set: it is non-empty since $I \neq \emptyset$; it is conjugacy closed since I is a union of H-conjugacy classes; and finally, since q is odd, if $g = us = su$ is the Jordan p-decomposition then $g^{\mathrm{ord}(g)/2} = s^{\mathrm{ord}(s)/2}$, and hence $g \in P(H, I)$ if and only if $s \in P(H, I)$.

To obtain the lower bound in Theorem 2.31 we used Theorem 2.30. A special subset $\mathscr{C}_0$ of F-conjugacy classes of W was examined, for which it was possible both to estimate $w_0 := | \cup_{C \in \mathscr{C}_0} C| / |W|$ and to find a good positive lower bound on $|T_C^F \cap P(H, I)|$ for each $C \in \mathscr{C}_0$. To give an understanding of this subset of W, while avoiding the technicalities associated with small dimensions and the other types of classical groups, we confine our attention to $H = \mathrm{GL}(n, q)$ with $n \geq 7$. Here $\mathscr{C}_0$ is a set of conjugacy classes in $W = S_n$. We choose a particular positive integer a as follows, and take $W_0 := \cup_{C \in \mathscr{C}_0} C$ to consist of all permutations with a single cycle of length $2^a k \in (n/3, 2n/3]$, for some integer k, and no other cycle of length divisible by 2^a. For $a_0 = \log_2 \ln 2 + \log_2 \log_2 n$, we take a to be the integer in the interval $[a_0 - 1/2, a_0 + 1/2)$. We note for later use that, since $n \geq 7$, we have $a \geq 1$ and $(13/4) \cdot 2^a \leq n$.

First we show that $|T_C^F \cap P(H, I)| / |T_C^F| \geq 1/2$, for $C \in \mathscr{C}_0$ with a cycle of length $2^a k$ as above. Each torus T_C^F in the H-conjugacy class of tori corresponding to C is of the form $Z \times A$, where Z is cyclic of order $q^{2^a k} - 1$ leaving invariant a subspace U of dimension $2^a k$ and acting as a Singer cycle on U, and for each $x \in A$, the 2-part of $\mathrm{ord}(x)$ (that is, the highest power of 2 dividing $\mathrm{ord}(x)$) is strictly less than the 2-part of $q^{2^a k} - 1$. Now half of the elements $z \in Z$ are such that the 2-part of $\mathrm{ord}(z)$ is equal to the 2-part of $q^{2^a k} - 1$, and for each such z, and any $x \in A$, the element zx has even order, and $(zx)^{|zx|/2}$ is the unique involution z_0 in Z. The element z_0 acts as $-I$ on the subspace U and has fixed point subspace of dimension $n - 2^a k \in [n/3, 2n/3)$; that is to say, z_0 is a strong involution and $zx \in \mathrm{P}(H, I)$. Thus $|T_C^F \cap \mathrm{P}(H, I)|/|T_C^F| \geq 1/2$.

Theorem 2.30 now implies that

$$\frac{|\mathrm{P}(H, I)|}{|H|} \geq \frac{1}{2} \cdot \frac{|W_0|}{|W|},$$

so it remains to estimate the size of W_0. A straightforward counting argument yields

$$\frac{|W_0|}{|W|} = \sum_k \frac{p_{\neg 2^a}(n - 2^a k)}{2^a k}, \tag{2.3}$$

where the sum is over integers k such that $n/3 < 2^a k \leq 2n/3$, and $p_{\neg 2^a}(n - 2^a k)$ is the proportion of elements in $S_{n-2^a k}$ with no cycle of length divisible by 2^a. By Lemma 4.2(a) of [58], which is based on Theorem 2.26,

$$p_{\neg 2^a}(n - 2^a k) > \frac{1}{4}(n - 2^a k)^{-1/2^a} > \frac{1}{4}n^{-1/2^a}.$$

Thus each summand in (2.3) is at least $3/(8n^{1+1/2^a})$ since $2^a k \leq 2n/3$. The number of summands in (2.3) is at least $(2n/3 - n/3)/2^a - 1 = n/(3 \cdot 2^a) - 1$, which is at least $n/(39 \cdot 2^a)$ (since $(13/4) \cdot 2^a \leq n$). Hence

$$\frac{|\mathrm{P}(H, I)|}{|H|} \geq \frac{1}{2} \cdot \frac{|W_0|}{|W|} \geq \frac{1}{2} \cdot \frac{n}{39 \cdot 2^a} \cdot \frac{3}{8n^{1+1/2^a}} = \frac{1}{208} \cdot \frac{1}{2^a \cdot n^{1/2^a}},$$

which is greater than $\frac{1}{208} \frac{1}{3\log_2(n)} = \frac{1}{624\log_2(n)}$. This proves Theorem 2.31 for $H = \mathrm{GL}(n, q)$.

The family W_0 of elements of the Weyl group W gives a far better lower bound than bounds obtained by geometric arguments. However we have not considered all conjugacy classes in W, and indeed it seems that, for this problem, application of "quokka theory" does not yield an upper bound. It is reasonable to ask how good the lower bound of Theorem 2.31 is. To attempt to answer this question, we quote a few sentences from [58, p. 3399].

> We did some numerical experiments for small $q \in \{3, 5, 9, 13\}$ and groups from the theorem up to dimension 1000. We computed many pseudo-random elements and checked if they powered up to an involution with a fixed point space of dimension in the right range. The proportion of these elements is not a monotonic function in the dimension, but the trend was that the proportion was about 25% for small dimensions and went down to about 15% in dimension 1000 (independently of the type of the group and q). Further, statistical tests on the data from the groups H we sampled strongly indicates that $\mathrm{P}(H, I)/|H| = O(1/\log(\ell))$. This seems to suggest at least that we cannot expect that there is a lower bound independent of the rank of the group.

2.3.6 Regular Semisimple Elements and Generating Functions

Let H be an n-dimensional classical group in odd characteristic, as in one of the lines of Table 2.1. The methods described in Sect. 2.3.4 show how to find a strong involution efficiently, or more generally, how to find an (α, β)-balanced involution z. The problem of constructing the centraliser $C_H(z)$ of such an involution will be discussed in Sect. 2.4. In this section we explore an estimation problem connected with part of the construction. An essential component in finding $C_H(z)$ is to take random conjugates z^g to find a "nice product" $y := zz^g$, where "nice" means "close to regular semisimple". This procedure is discussed in the seminal paper [78] by Christopher Parker and Rob Wilson. They estimate that $O(n)$ random products will produce a nice product with high probability. The approach taken by Praeger and Seress [86], and described in this section, shows that only $O(\log n)$ random products are required.

Written in an appropriate basis, the product $y = zz^g$ of an involution z and a random conjugate z^g of z has the following form, where y_0 has no ± 1-eigenvectors:

$$
y := zz^g = \begin{pmatrix} I_r & 0 & 0 \\ 0 & y_0 & 0 \\ 0 & 0 & -I_s \end{pmatrix}.
$$

Typically, the dimension r is close to $2m - n$, where m is the maximum of the dimensions of the ± 1-eigenspaces of z, and s is close to 0. The question arises: what kind of matrix do we expect for y_0 "typically"? Let us restrict attention to the simplest case where $H = GL(n, q)$ with q odd. By considering the results of computer experiments on various (α, β)-balanced involutions and their random conjugates for various n and odd q, we discovered that often y_0 is "regular semisimple". *For the following discussion, let us assume that $y = y_0$.*

An element y of $GL(n, q)$ is called *semisimple* if is diagonalisable over some extension field of $\mathbb{F}_q$ (see [17, p. 11]), and this is equivalent to its minimal polynomial $m_y(t)$ being multiplicity free. Also y is called *regular* if its centraliser in the corresponding general linear group over the algebraic closure of $\mathbb{F}_q$ has minimal possible dimension, namely n (see [17, p. 29]). It turns out that an element y of a general linear group is regular if and only if $m_y(t) = c_y(t)$, where $c_y(t)$ denotes the characteristic polynomial of y. These two conditions for elements of finite classical groups are discussed and compared in [69, Note 8.1]. The *regular semisimple elements* are those which are both regular and semisimple. In fact, for elements y of $H = GL(n, q)$, y is regular semisimple if and only if the characteristic polynomial $c_y(t)$ for its action on $V(n, q)$ satisfies

$$c_y(t) = \text{a product of pairwise distinct irreducible polynomials.}$$

Looking into the analysis of this situation in the paper [78], it is clear that Parker and Wilson recognised that regular semisimple elements y occur frequently. Moreover, the proportion of regular semisimple elements in the full n-dimensional matrix algebra was estimated by Neumann and Praeger [70]. The main result of [86] (cf. Theorem 2.34) is a strengthening of the estimates in [70, 78].

The characteristic polynomial $c_y(t)$ has two special properties: firstly, when $y = y_0$ the element y has no ± 1-eigenvectors, so $c_y(t)$ is not divisible by $t \pm 1$. Secondly, since $y^z = z^{-1}(zz^g)z = z^g z = y^{-1}$, the characteristic polynomials of y and y^{-1} are equal. Now $c_{y^{-1}}(t) = c_y^*(t)$ is the conjugate polynomial of $c_y(t)$ where, for an arbitrary polynomial $f(t)$ with $f(0) \neq 0$, its conjugate polynomial is $f^*(t) := f(0)^{-1} t^{\deg f} f(t^{-1})$. Thus $c_y(t) = c_y^*(t)$ is self-conjugate. We have seen that conjugation by z inverts y, and similarly conjugation by z^g inverts y. Inverting a regular semisimple matrix pins down the conjugacy class of the involution z, as shown in [86, Lemma 3.1]. For n even and q odd, let $\mathscr{C} \subseteq \mathrm{GL}(n, q)$ denote the the conjugacy class of involutions with fixed point space of dimension $n/2$.

Lemma 2.33. *Let $z, y \in \mathrm{GL}(n, q)$ with q odd, such that y is regular semisimple with characteristic polynomial $c_y(t)$ coprime to $t^2 - 1$, and z is an involution inverting y. Then n is even, $z \in \mathscr{C}$, and zy is also an involution which inverts y.*

By Lemma 2.33, we have a bijection $(z', z) \mapsto (y, z)$ between the sets

$$ X = \left\{ (z, z') \in \mathscr{C} \times \mathscr{C} \;\middle|\; \begin{array}{l} y := zz' \text{ regular semisimple} \\ \text{with } c_y(t) \text{ coprime to } t^2 - 1 \end{array} \right\} $$

and

$$ Y = \left\{ (y, z) \;\middle|\; \begin{array}{l} y, z \in \mathrm{GL}(n, q), z^2 = 1, y^z = y^{-1} \\ y \text{ regular semisimple, and} \\ c_y(t) \text{ coprime to } t^2 - 1 \end{array} \right\}. $$

The set X is relevant for algorithmic purposes, while the set Y is more amenable to estimation techniques. For the algorithm, we are given (that is to say, we have already found) the involution $z \in \mathscr{C}$, and we want to know the proportion of $z' \in \mathscr{C}$ such that $(z, z') \in X$. This is

$$ \frac{|\{z' \in \mathscr{C} \mid (z, z') \in X\}|}{|\mathscr{C}|} = \frac{|X|}{|\mathscr{C}|^2} = \frac{|Y|}{|\mathscr{C}|^2} = \frac{|\mathrm{GL}(n, q)|}{|\mathscr{C}|^2} \cdot \frac{|Y|}{|\mathrm{GL}(n, q)|} $$

and the first factor on the right of the equality, namely $\frac{|\mathrm{GL}(n,q)|}{|\mathscr{C}|^2} = \frac{|\mathrm{GL}(n/2,q)|^4}{|\mathrm{GL}(n,q)|}$, lies between $(1 - q^{-1})^7$ and $(1 - q^{-1})^2$. Thus the essential problem is to estimate

$$ ss(n, q) := \frac{|Y|}{|\mathrm{GL}(n, q)|}. $$

Parker and Wilson [78] give a heuristic that estimates this quantity as being at least c/n if we require in addition that y has odd order. Our approach gives a surprisingly precise answer; see [86, Theorem 1.2]. Since n is even we consider $ss(2d, q)$.

Theorem 2.34. *For a fixed odd prime power q, the limit of $ss(2d,q)$ as $d \to \infty$ exists and*

$$ss(\infty, q) := \lim_{d \to \infty} ss(2d, q) = (1 - q^{-1})^2.$$

Moreover $|ss(2d,q) - ss(\infty,q)| = o(q_0^{-d})$ for any q_0 such that $1 < q_0 < \sqrt{q}$.

Corollary 2.35. *There exists $c > 0$ with the property that for any $z \in \mathscr{C}$ the proportion of $z' \in \mathscr{C}$ such that $(z, z') \in X$ is bounded below by c.*

We use generating functions discussed in Sect. 2.2.5 to study the quantities $ss(2d, q)$. We define

$$S(u) = \sum_{d=0}^{\infty} ss(2d, q) u^d \quad \text{where} \quad ss(0, q) = 1.$$

Since y is regular semisimple, $c_y(t)$ is multiplicity-free, and since y is inverted by the involution z, we have a factorisation

$$c_y(t) = \left(\prod_{i=1}^{r} f_i(t) \right) \times \left(\prod_{j=1}^{s} g_j(t) g_j^*(t) \right) \tag{2.4}$$

where each $f_i = f_i^*$ has even degree, and each $g_j \neq g_j^*$, with the f_i, g_j, g_j^* pairwise distinct monic irreducibles. We use this decomposition to find in [86, Lemma 3.2] that the number of pairs $(y', z) \in Y$ such that y' has characteristic polynomial $c_y(t)$ is equal to

$$\frac{|\mathrm{GL}(2d, q)|}{\left(\prod_{i=1}^{r} (q^{\frac{1}{2} \deg f_i} - 1) \right) \left(\prod_{j=1}^{s} (q^{\deg g_j} - 1) \right)}.$$

Summing over all possible $c_y(t)$ gives an expression for $ss(2d, q) |\mathrm{GL}(2d, q)|$. Comparing the expression we obtain for $ss(2d, q)$ by this process with the coefficient of u^d in the infinite product

$$\prod_{f = f^*, \text{ irred.}} \left(1 + \frac{u^{\frac{1}{2} \deg f}}{q^{\frac{1}{2} \deg f} - 1} \right) \times \prod_{\{g, g^*\}, g \neq g^*, \text{ irred.}} \left(1 + \frac{u^{\deg g}}{q^{\deg g} - 1} \right),$$

we see that the two expressions are the same. Hence $S(u)$ is equal to this infinite product. The contribution to the infinite product from each irreducible polynomial f or conjugate pair $\{g, g^*\}$ of non-self-conjugate polynomials depends only on the degrees of the polynomials. Thus

$$S(u) = \prod_{m \geq 1} \left(1 + \frac{u^m}{q^m - 1} \right)^{N^*(q; 2m)} \times \prod_{m \geq 1} \left(1 + \frac{u^m}{q^m - 1} \right)^{M^*(q; m)} \tag{2.5}$$

where the exponents are

$$N^*(q;m) = \#\text{monic irreducible self-conjugate polynomials over}$$
$$\mathbb{F}_q \text{ of degree } m.$$
$$M^*(q;m) = \#\text{ (unordered) conjugate pairs of monic irreducible}$$
$$\text{non-self-conjugate polynomials over } \mathbb{F}_q \text{ of degree } m.$$

It turned out that a somewhat similar infinite product arose when Praeger was studying separable matrices in finite unitary groups with Jason Fulman and Peter Neumann in [36]. A similar analysis to that given in [36] for these matrices yielded:

1. $S(u)$ is analytic for $|u| < 1$ with a simple pole at $u = 1$.
2. $S(u) = (1 - u)^{-1} H(u)$, with $H(u)$ analytic for $|u| < \sqrt{q}$.

Completing the analysis we found the asymptotic behaviour of the $ss(2d, q)$, as in Theorem 2.34.

2.4 Computing Centralisers of Involutions

The results in the previous section play a significant role in the analysis of algorithms to compute centralisers of involutions. In general the problem of computing centralisers is of great importance in theoretical computer science and in group theory. In computer science, the main interest stems from the connection with the *graph isomorphism problem*.

Problem 2.36. *(ISO)* Given: graphs $\Gamma_1(V_1, E_1)$ and $\Gamma_2(V_2, E_2)$.
Find: an edge-preserving bijection between V_1 and V_2, or prove that no such bijection exists.

ISO is polynomial-time reducible to the following computational problems with permutation groups.

Problem 2.37. *(STAB)* Given: a permutation group $G \leq \text{Sym}(\Omega)$ and a subset $\Delta \subseteq \Omega$.
Find: the set stabiliser $\text{Stab}_G(\Delta) = \{g \in G \mid \Delta^g = \Delta\}$.

Problem 2.38. *(INT)* Given: permutation groups $G, H \leq \text{Sym}(\Omega)$.
Find: the intersection $G \cap H$.

Problem 2.39. *(CENT)* Given: permutation groups $G, H \leq \text{Sym}(\Omega)$.
Find: the centraliser $C_G(H) = \{g \in G \mid h^g = h \text{ for all } h \in H\}$.

Problems 2.37–2.39 are in the same class of the complexity hierarchy, which means that they can be reduced to each other in time polynomial in the input length [59].

The reduction of ISO is easiest to STAB or INT. First, we notice that $\Gamma_1(V_1, E_1)$ and $\Gamma_2(V_2, E_2)$ are isomorphic if and only if $\Gamma_1 \cup \Gamma_2$ (disjoint copies of Γ_1 and Γ_2) has an automorphism that exchanges V_1 and V_2. Therefore, it is enough to compute automorphism groups of graphs. Given a graph $\Gamma(V, E)$, define Ω as the set of unordered pairs in V. Then E corresponds to a subset $\Delta \subseteq \Omega$, and $\mathrm{Sym}(V)$ acts as a group G on Ω. We can compute $\mathrm{Aut}(\Gamma)$ as $\mathrm{Aut}(\Gamma) = \mathrm{Stab}_G(\Delta)$ or $\mathrm{Aut}(\Gamma) = G \cap (\mathrm{Sym}(\Delta) \times \mathrm{Sym}(\Omega \setminus \Delta))$.

Although, using backtrack methods (see e.g. [87, Chap. 9]), ISO and CENT are usually easy to solve in practice, no polynomial-time solution is known for Problems 2.36–2.39. Special cases with polynomial-time solutions are of great theoretical and practical interest.

In group theory, the most important case of centraliser computations is to construct *centralisers of involutions*. On the theoretical side, a major tool in the study and classification of finite simple groups is the investigation of their involution centralisers [41]. On the computational side, in the last decade involution centraliser computations became prevalent [1, 7, 45, 53, 56, 78]. In the next subsections, we describe some applications of centraliser computations; Bray's algorithm [16] for computing centralisers of involutions; and efforts to analyze Bray's algorithm.

2.4.1 Applications of Centralisers of Involutions Computations

A recent active area of computational group theory is the so-called *matrix group recognition project*. Let V be a finite dimensional vector space over a finite field $\mathbb{F}_q$. Given $G = \langle S \rangle \leq \mathrm{GL}(V)$, the goal is to compute quantitative and structural information about G such as the order, a composition series, and important characteristic subgroups like the largest solvable normal subgroup of G.

There are two main approaches to matrix group recognition. The *geometric approach*, initiated by Neumann and Praeger [69] and currently led by Leedham-Green and O'Brien [52,77], is based on Aschbacher's classification of matrix groups [2]. Aschbacher defines nine categories of matrix groups G. In seven of these categories, there is a natural normal subgroup $N \lhd G$ that can be used to divide the recognition problem into two smaller subproblems on N and G/N. Based on that result, the geometric approach tries to find a homomorphism $\varphi : G \to H$ into an appropriate permutation or matrix group H, and recursively recognise $\mathrm{Im}(\varphi)$ and $\mathrm{Ker}(\varphi)$. In contrast, the *black-box group approach* of Babai and Beals [4] aims for the abstract group theoretic structure of G. Babai and Beals define a series of characteristic subgroups, present in all finite groups, and initiate a program that tries to compute a composition series going through these characteristic subgroups.

Both approaches eventually lead to simple (or quasisimple) matrix groups, where further divide-and-conquer is impossible. For such groups, a major issue is the solution of the *constructive membership problem*.

2.4.2 Constructive Membership in Lie Type Groups

Definition 2.40. A *black-box group* G is a group whose elements are encoded by bit strings (strings consisting of 0s and 1s) of uniform length. Moreover, there are oracles for the following tasks. Given strings representing $g, h \in G$, we can compute a string representing gh; a string for g^{-1}; and we can decide whether $g = 1$.

A *black-box algorithm* is an algorithm that, given G by a set of generators, uses only the black-box oracles.

The definition of black-box groups covers the "concrete" representations of groups as permutation groups or matrix groups defined over finite fields. Note that if G is a black-box group and N is a *recognisable* normal subgroup (i.e., given a string representing some $g \in G$, we can decide whether $g \in N$), then G/N is also a black-box group. This observation plays a crucial role in recursive algorithms, allowing us to work in factor groups. Also note that we require only that N is recognisable, but N is not necessarily *constructed* (i.e., we may not have a generating set for N in hand). Examples of recognisable normal subgroups that may be hard to construct are the centre and the largest soluble normal subgroup of G. Black-box groups were introduced by Babai and Szemerédi [5]. For an introduction to the basic black-box group algorithms, see [87, Chap. 2].

A black-box group algorithm does not use specific features of the group representation, nor particulars of how group operations are performed. For example, we lose all information stored implicitly in the cycle structure of a permutation, or in the characteristic polynomial of a matrix. In practice, and also in some theoretical considerations, we often allow oracles for some other operations; an example is an oracle to compute element orders.

The very reasonable and justified question arises: why do we handicap ourselves with black-box group algorithms? One answer is that in certain situations, we cannot do more than the black-box operations. For example, to generate random elements in a matrix group, so far every algorithm takes repeated products and inverses of the given generators, and after a while declares the last element constructed as a random element of the input group [3, 18, 24]. Bray's algorithm (see Sect. 2.4.4) for computing centralisers of involutions is another example of a black-box group algorithm, with a possible enhancement using element order oracles. Another, more unusual answer is that elements of a permutation group can be described as unique words in a strong generating set (SGS), constructed in a canonical way. The group operations are performed using the images of elements of the base associated with the SGS. For the important class of small-base groups, these group operations are much faster than permutation multiplication, but the algorithms using this representation are strictly black-box. For details, we refer to [87, Chap. 5.4].

Next, we define the notion of a straight-line program (SLP). Expressing elements of a group G in a given set of generators may result in words of length proportional to $|G|$; intuitively, SLPs are shortcuts, to reach group elements faster from a set of generators. By [5], every $g \in G$ can be reached from any set of generators by an SLP of length at most $(1 + \log |G|)^2$.

Definition 2.41. Given $G = \langle S \rangle$ and $g \in G$, a *straight-line program (SLP)* reaching g from S is a sequence of expressions $W = (w_1, \ldots, w_m)$ such that, for $i = 1, 2, \ldots, m$,

1. w_i is a symbol for some $s \in S$; or
2. $w_i = (w_j, w_k)$ for some $j, k < i$; or
3. $w_i = (w_j, -1)$ for some $j < i$.

We define the evaluation of W the natural way: $\mathrm{eval}(w_j, w_k) = \mathrm{eval}(w_j)\mathrm{eval}(w_k)$ and $\mathrm{eval}(w_j, -1) = \mathrm{eval}(w_j)^{-1}$; and require that $\mathrm{eval}(w_m) = g$.

Finally, we are ready to define the constructive membership problem.

Definition 2.42. A *constructive membership algorithm for a group G* is a black-box group algorithm that, given the black-box group $G = \langle S \rangle$ and $g \in G$, constructs an SLP reaching g from S.

The main result of this subsection is the following theorem by Holmes et al. [45].

Theorem 2.43 ([45]). *Let G be a black-box group equipped with an order oracle. There is a black-box Monte Carlo algorithm which reduces the constructive membership problem for G to three instances of the same problem for centralisers of involutions of G.*

Proof. Let $G = \langle S \rangle$ and $g \in G$. An algorithm constructing an SLP reaching g from S consists of the following steps.

1. Find $h \in G$ with $\mathrm{ord}(gh) = 2\ell$. Define $z := (gh)^\ell$.
2. Find an involution $x \in G$ with $\mathrm{ord}(xz) = 2m$. Define $y := (xz)^m$.
3. Construct $X = C_G(x)$.
4. Solve the constructive membership problem for $y \in X$.
5. Construct $Y = C_G(y)$.
6. Solve the constructive membership problem for $z \in Y$.
7. Construct $Z = C_G(z)$.
8. Solve the constructive membership problem for $gh \in Z$.
9. Compute and return an SLP for g.

To prove the correctness of the algorithm, observe that z, constructed in Step 1, is an involution centralising gh. In Step 2, y is in the centre of the dihedral group $\langle x, z \rangle$, so x is an involution centralising y and y is an involution centralising z. Hence Steps 3, 5 and 7 compute centralisers of involutions, and the constructive membership problems in Steps 4, 6 and 8 indeed try to reach elements of G that are in the appropriate subgroups. Finally, note that the construction of x provides an SLP reaching x from S and, consequently, we have SLPs reaching y, then z, then gh from S. Also, in Step 1, we construct an SLP reaching h from S. Hence, in Step 9, we can construct an SLP reaching g from S. □

Remark 2.44. We note that the hypothesis of Theorem 2.43 that G has an order oracle can be relaxed. The only places in the algorithm where the order oracle is used are in Steps 1 and 2. For example, at the construction of z in Step 1, we can proceed the following way. Instead of computing ℓ, we can raise gh to an appropriate multiple of the odd part of $|G|$. To find such a multiple (without knowing $|G|$), it is enough to know a superset of primes occurring in $|G|$ or, in the case of a matrix group $G \leq \mathrm{GL}(n, q)$, we can work with the set of *pseudoprimes*: these are the largest divisors of the numbers $q^e - 1$ for $e \leq n$, that are relatively prime to $q^j - 1$ for all $j < e$. The pseudoprimes can be computed in polynomial time (polynomial in terms of n and $\log q$). For details, see [4]. The use of the order oracle in Step 2 can be avoided in exactly the same way.

In [45], Holmes et al. show that if G is a simple group of Lie type then the algorithm described in Theorem 2.43, not counting the time requirement of Steps 4, 6 and 8, runs in polynomial time. However, we cannot apply the theorem recursively to the groups in these steps, because they are not simple. Therefore, we need a recursive scheme involving all groups, not only the simple ones. Such a scheme is designed by Babai et al. in [7]; Theorem 2.43 is a crucial ingredient in the following result.

Theorem 2.45 ([7]). *There is a randomised polynomial-time algorithm, employing certain number-theoretical oracles, which, given a matrix group $G \leq \mathrm{GL}(n, q)$ of odd characteristic, solves the constructive membership problem in G.*

The required *number-theoretical oracles* are the *factorisation of integers* of the form $q^e - 1$, for $1 \leq e \leq n$, and the *solution of the discrete logarithm problem*: given $a, b \in \mathbb{F}_{q^e}^*$, decide whether $a \in \langle b \rangle$; and, if the answer is affirmative, then find an integer x such that $a = b^x$. In polynomial-time algorithms for matrix groups, it is customary to assume the use of these number theory oracles as they are already needed in finding a composition series and the order of a 1×1 matrix group over $\mathbb{F}_q$. We note that Theorem 2.45 extends to matrix groups defined over fields of characteristic 2, with some restrictions on the composition factors of G. It is expected that these restrictions will be removed in the near future, as constructive membership algorithms in all simple groups are in the offing.

2.4.3   Constructive Recognition of Lie Type Groups

Membership testing is an important first step in exploring a permutation or matrix group G; however, for studying the structure of G and constructing important subgroups, it is beneficial to identify the composition factors of G with standard copies of these factor groups. For alternating and classical groups, the standard copy is the natural permutation and matrix representation, respectively. For exceptional groups, the definition of a standard copy is not so clear-cut: we may choose the smallest-dimensional matrix representation, or a Bruhat decomposition, or any other representation we may be able to control. Here we only give a formal definition for classical groups, taken from [48].

Definition 2.46. *Constructive recognition* of a black-box group $G = \langle S \rangle$ isomorphic to a simple classical group defined on some vector space over a field of given characteristic p is an algorithm that verifies that there is, indeed, an isomorphism, and finds the following:

(i) The field size $q = p^e$, as well as the type and the dimension d of G.
(ii) A new set S^* generating G, a vector space $\mathbb{F}_q^d$, and a monomorphism $\lambda \colon G \to$ PSL(d, q), specified by the image of S^*, such that $G\lambda$ acts projectively on $\mathbb{F}_q^d$ as a classical group defined on $\mathbb{F}_q^d$.

Moreover, the data structures underlying (ii) yield deterministic algorithms for each of the following:

(iii) Given $g \in G$, find $g\lambda$ and a straight-line program from S^* to g.
(iv) Given $h \in$ PGL(d, q), decide whether or not $h \in G\lambda$; and, if it is, find $h\lambda^{-1}$ and a straight-line program from S^* to $h\lambda^{-1}$.
(v) Find a form on $\mathbb{F}_q^d$ involved in the definition of G as a classical group, if $G \ncong$ PSL(d, q).

Although Definition 2.46 is formulated in the general context of black-box groups, of course it can be applied to any given permutation or matrix representation of G. The simplest but most important case is when G is already given in its natural representation, and the only task is to find "nice" generators S^* such that each element of G can be reached easily from S^*. For classical groups of odd characteristic, this task has been accomplished by Leedham-Green and O'Brien by a highly efficient algorithm [53]. A rough outline of their procedure is given in Algorithm 5.

Algorithm 5: CONSTRUCTIVERECOGNITION

Input: $G = \langle S \rangle \leq$ GL$(V) \cong$ GL(n, q), q odd, G is a classical group in its natural
      representation;
Output: A data structure for constructive recognition of G;
(1) **repeat**
    $y :=$ random element of G;
until ord(y) *is even and* $x := y^{\text{ord}(y)/2}$ *has* ± 1*-eigenspaces* E_1, E_{-1} *with*
dim$(E_i) \in (n/3, 2n/3)$;
(2) Construct $H = C_G(x)$;
(3) Recursively solve constructive recognition for the restriction of H to its action on E_1
and E_{-1};
(4) Use the result of Step (3) to obtain nice generators and data structure for constructive
recognition of G;

The following simple lemma from [84] implies that Step (3) is indeed a recursive call.

Lemma 2.47. *Let G, x, E_1, E_{-1} be as in Algorithm 5, with G classical but not linear. Then $V = E_1 \perp E_{-1}$, and both E_1 and E_{-1} are nondegenerate (and of even dimension if G is symplectic).*

Proof. For $u \in E_1$ and $w \in E_{-1}$, we have $(u, w) = (u, w)^x = (u, -w)$ and hence $(u, w) = 0$. Thus $E_1 \subseteq E_{-1}^\perp$. Since the bilinear form is nondegenerate, $\dim(E_1) = n - \dim(E_{-1}) = \dim(E_{-1}^\perp)$ and hence $E_1 = E_{-1}^\perp$. Therefore, $E_{-1} \cap E_{-1}^\perp = 0$ so E_{-1}, and similarly also E_1, are nondegenerate. In particular, E_1 and E_{-1} both have even dimension if G is symplectic. $\square$

Since, for $i \in \{1, -1\}$, x acts as a scalar matrix on E_i, Lemma 2.47 implies that the restriction of H to E_i is a classical group of the same type as G and Step (3) is indeed a recursive call. Note that the requirement $\dim(E_i) \in (n/3, 2n/3)$ ensures that $C_G(x)$ can be split into two parts of roughly equal size, thereby ensuring that the depth of the recursion is logarithmic in n.

To analyze Algorithm 5, for the first two steps we have to estimate (i) the proportion of elements y as in Step (1); and (ii) give a running time estimate for the construction of involution centralisers. Task (i) has been accomplished in Sect. 2.3.4. In the next two subsections, we describe and analyze an algorithm for computing involution centralisers.

2.4.4 Computation of an Element Centralising an Involution

In this subsection we describe an algorithm by Bray [16] that constructs an element in the centraliser of a given involution.

Algorithm 6: CENTRALISINGELEMENT

Input: $G = \langle S \rangle$ and an involution $x \in G$;
Output: An element of $C_G(x)$;
(1) $g :=$ random element of G;
(2) $y := x \cdot x^g$ and $m := \mathrm{ord}(y)$;
(3) **if** *m is even* **then**
    return $\zeta(g) := y^{m/2}$
else
    return $\zeta(g) := y^{(m+1)/2} g^{-1}$
end

We note that the order computation in Step (2) may be avoided, using a superset of primes occurring in G, or pseudoprimes (see Remark 2.44).

Lemma 2.48. *The output of Algorithm 6 is correct: no matter which $g \in G$ is chosen in Step (1), we have $\zeta(g) \in C_G(x)$.*

Proof. For any $g \in G$, the group $D := \langle x, x^g \rangle$ is dihedral, of order $2m$. If m is even then $\zeta(g) \in Z(D)$; in particular, $\zeta(g)$ centralises $x \in D$.

If m is odd then, using that $x^2 = 1$, we obtain

$$x^{y^{\frac{m+1}{2}}} = (xg^{-1}xg)^{\frac{m-1}{2}} x (xg^{-1}xg)^{\frac{m+1}{2}} = x^g.$$

Comparison of the leftmost and rightmost terms gives $\zeta(g) = y^{\frac{m+1}{2}} g^{-1} \in C_G(x)$.
□

We say that $g \in G$ is of *even type* if $y = xx^g$ has even order, and $g \in G$ is of *odd type* if $y = xx^g$ has odd order. Note that for any $c \in C_G(x)$, $x^{cg} = x^g$, so $xx^g = xx^{cg}$ and consequently g and cg have the same type. Moreover, $(xx^g)^c = xx^{gc}$ so xx^g and xx^{gc} are conjugate, have the same order, and g and gc have the same type. Combining the last two observations, we obtain that *in a double coset $C_G(x) \cdot g \cdot C_G(x)$, all elements have the same type.*

Lemma 2.49. (i) *If g is chosen from the uniform distribution on the set of odd type elements of G then $\zeta(g)$ is a uniformly distributed random element of $C_G(x)$.*

(ii) *If g is chosen from the uniform distribution on the set of even type elements of G and $\zeta(g)$ is in the conjugacy class $\mathscr{C}$ of involutions in $C_G(x)$ then $\zeta(g)$ is a uniformly distributed random element of $\mathscr{C}$.*

Proof. (i) Suppose that g is of odd type. For $c \in C_G(x)$, we have $y^{\frac{m+1}{2}}(cg)^{-1} = y^{\frac{m+1}{2}} g^{-1} c^{-1}$ and so $\zeta(cg) = \zeta(g)c^{-1}$. Hence, as cg runs through the coset $C_G(x) \cdot g$, $y^{\frac{m+1}{2}} g^{-1} c^{-1}$ runs through $C_G(x)$. This implies that if g runs through the elements of G of odd type then each element of $C_G(x)$ occurs as $\zeta(g)$ exactly the same number of times.

(ii) Suppose now that g is of even type. Then $\zeta(g) = (xx^g)^{m/2}$ is an involution; let $\mathscr{C}$ denote its conjugacy class in $C_G(x)$. As gc runs through the coset $g \cdot C_G(x)$, $\zeta(gc) = (xx^{gc})^{m/2} = ((xx^g)^{m/2})^c$ covers each element of $\mathscr{C}$ the same number of times. Hence each element of a fixed conjugacy class $\mathscr{C}$ of involutions in $C_G(x)$ has the same chance to occur as $\zeta(g)$ for some g of even type.   □

2.4.5 Computation of the Full Centraliser

In order to compute a set X of generators of $C_G(x)$ for a given group G and involution $x \in G$, we may construct a sequence $(g_1, \ldots, g_m)$ of random elements in G and take $X := \{\zeta(g_i) \mid 1 \leq i \leq m\}$. By Lemma 2.48, we always have $\langle X \rangle \leq C_G(x)$, but when can we stop? How large should m be so that, with high probability, X generates the entire group $C_G(x)$?

By Lemma 2.49, random elements g_i of odd type are highly desirable, since then $\zeta(g_i)$ is a uniformly distributed random element of $C_G(x)$. Such a random element

$\zeta(g_i) \in C_G(x)$, added to an already constructed proper subgroup $H < C_G(x)$, increases H with probability $1 - 1/|C_G(x) : H| \geq 1/2$, so if we know an upper bound ℓ for the length of subgroup chains in $C_G(x)$ then we may estimate how many elements g_i of odd type we need to encounter. For polynomial-time computations, the trivial bound $\ell \leq \log_2 |G|$ suffices, but sometimes we have much better estimates for the number of required random generators. In particular, in the especially important case when G is a simple group of Lie type defined over a field of odd characteristic, the structure of involution centralisers is known. Consequently, for any involution $x \in G$, the number of uniformly distributed random elements needed to generate $C_G(x)$ with probability greater than $1 - \varepsilon$ can be bounded by a function of ε, independent of G and x [57]. Therefore, the following seminal result of Parker and Wilson [78] has great importance in the analysis of many matrix group algorithms.

Theorem 2.50 ([78]). *There exists a positive constant c such that:*

(i) *If G is a simple exceptional group of Lie type defined over a field of odd order, and x is any involution in G, then the probability that a uniformly distributed random element $g \in G$ is of odd type is bounded below by c.*

(ii) *If G is a simple classical group defined over a field of odd order, with natural module of dimension n, and x is any involution in G, then the probability that a uniformly distributed random element $g \in G$ is of odd type is bounded below by c/n. Moreover, the order of magnitude $1/n$ for a lower bound is best possible.*

Parker and Wilson [78, p. 886] give an indication of how big the constants can be: "The constants c that can be obtained from our proofs are of the order of $1/1000$, but we have made no attempt to calculate them explicitly, as we conjecture that the best possible constants are nearer $1/4$."

The basic idea of the proof of Theorem 2.50 is to identify a set of dihedral subgroups D of twice odd order in G, each D containing the given involution x. If the random conjugate x^g falls into one of these subgroups D then xx^g has odd order and g is of odd type. In order to avoid double counting, we also require that generators of the maximal cyclic normal subgroup of D be *regular semisimple* in a suitable subgroup $H \leq G$. (Here H depends on D but H is also of Lie type. We require the generators of D to be regular semisimple as elements of this Lie type group, as defined in Sect. 2.3.6.)

While Theorem 2.50 is sufficient to prove polynomial running time of centraliser of involution computations in Lie type simple groups, the scarcity of elements of odd type raises the the following questions. Is there an algorithm that uses the lower quality random elements $\zeta(g_i) \in C_G(x)$, obtained from g_i of even type, to generate $C_G(x)$? Can the asymptotic running time of this algorithm be faster than the construction of $C_G(x)$ using the uniformly distributed $\zeta(g_i)$ obtained from g_i of odd type? To formulate this problem precisely, we need some definitions.

We consider finite classical groups H of dimension n over a finite field $\mathbb{F}_q$ of odd order q. We denote by H^* the generalized Fitting subgroup of H (for example $H^* = \mathrm{SL}(n, q)$ if $H = \mathrm{GL}(n, q)$). Let α, β be real numbers such that $0 < \alpha <$

$1/2 < \beta < 1$, and let $x \in H$ be of order 2. Recall that x is called an (α, β)-*balanced involution* in H if the subspace $E_1(x)$ of fixed points of x in the underlying vector space has dimension r where $\alpha n \leq r < \beta n$. For a given sequence $\mathcal{X} = (\mathcal{C}_1, \ldots, \mathcal{C}_m)$ of conjugacy classes of (α, β)-balanced involutions in H, a c-tuple $(g_1, \ldots, g_m)$ is a *class-random sequence from* $\mathcal{X}$ if g_i is a uniformly distributed random element of $\mathcal{C}_i$ for each $i = 1, \ldots, m$, and the g_i are mutually independent.

Given a classical group $G \leq GL(n, q)$ and an involution $x \in G$, the centraliser $C_G(x)$ modulo x is the direct product of two classical groups $H^{(1)}$ and $H^{(-1)}$, acting on $E_1(x)$ and $E_{-1}(x)$, respectively. If $g \in G$ is of even type then $\zeta(g)$ acts as an involution $g^{(J)}$ on E_J, for $J \in \{1, -1\}$, and if $(g_1, \ldots, g_m)$ is a sequence of uniformly distributed random elements of even type in G then Lemma 2.49 implies that $(g_1^{(J)}, \ldots, g_m^{(J)})$ is a class-random sequence from some conjugacy classes of involutions $\mathcal{X}^{(J)} = (\mathcal{C}_1^{(J)}, \ldots, \mathcal{C}_m^{(J)})$.

With an application in Algorithm 5 in mind, we propose the following problems. We use the notation and definitions of the previous paragraphs.

Problem 2.51. Given a classical group $G \leq GL(n, q)$ and a $(1/3, 2/3)$-balanced involution x in G, estimate the probability p that for a uniformly distributed $g \in G$ of even type, $g^{(J)}$ is an (α, β)-balanced involution in $H^{(J)}$, for *both* $J \in \{1, -1\}$. Here α, β are constants, chosen appropriately.

Problem 2.52. Let $G \leq GL(n, q)$ be a classical group and let $\mathcal{X} = (\mathcal{C}_1, \ldots, \mathcal{C}_m)$ be a sequence of conjugacy classes of (α, β)-balanced involutions in G. Estimate the minimum value of m such that, with high probability, a class-random sequence from $\mathcal{X}$ generates a subgroup of G containing G^*.

If the product $(1/p)m$, for the probability p from Problem 2.51 and the minimum value m from Problem 2.52, satisfies $(1/p)m = o(n)$ then the elements $\zeta(g)$ obtained from even type g generate $C_G(x)$ asymptotically faster than the elements $\zeta(g)$ obtained from odd type g.

Problem 2.52 has been solved for all classical groups.

Theorem 2.53 ([84]). *Let α, β be real numbers such that $0 < \alpha < 1/2 < \beta < 1$. Then there exist integers $m = m(\alpha, \beta)$ and $n(\alpha, \beta)$ such that, for G, n, q as above, with q odd, if $n > n(\alpha, \beta)$ and $\mathcal{X} = (\mathcal{C}_1, \ldots, \mathcal{C}_m)$ is a given sequence of conjugacy classes of (α, β)-balanced involutions in G, then a class-random sequence from $\mathcal{X}$ generates a subgroup containing G^* with probability at least $1 - q^{-n}$.*

The basic idea of the proof of Theorem 2.53 is standard: if a class-random sequence $(g_1, \ldots, g_m)$ does not generate G^* then all g_i belong to some maximal subgroup $M < G$, with M not containing G^*. Since g_i is uniformly distributed in its conjugacy class, we have to estimate the ratios $|M \cap \mathcal{C}_i|/|\mathcal{C}_i|$ for all maximal subgroups M. Maximal subgroups are characterised by Aschbacher's theorem [2]; it turns out that the most difficult case is when M is reducible (has a proper invariant subspace).

Much less is known about Problem 2.51. At present, a solution is known only in the case when $G^* = SL(n, q)$.

Theorem 2.54 ([85]). *There exist c and n_0 such that if $n > n_0$, $\mathrm{SL}(n,q) \leq G \leq \mathrm{GL}(n,q)$, x is a $(1/3,2/3)$-balanced involution of G, and $g \in G$ is a uniformly distributed random element among the elements of G of even type, then with probability at least $c/\log n$, $g^{(1)}$ and $g^{(-1)}$ are $(1/6,2/3)$-balanced involutions on the eigenspaces $E_1(x)$ and $E_{-1}(x)$ respectively.*

The proof of Theorem 2.54 uses a significant enhancement of the generating function method described in Sect. 2.3.6, and also some ideas from [58].

Acknowledgements This chapter forms part of our Australian Research Council Discovery Project DP110101153. Praeger and Seress are supported by an Australian Research Council Federation Fellowship and Professorial Fellowship, respectively. Niemeyer thanks the Lehrstuhl D für Mathematik at RWTH Aachen for their hospitality, and acknowledges a DFG grant in SPP1489. All three of us warmly thank the de Brún Centre for Computational Algebra at National University of Ireland, Galway, for their hospitality during the Workshop on Groups, Combinatorics and Computing in April 2011, where we presented the short lecture course that led to the development of this chapter. We are very grateful to Peter M. Neumann for many thoughtful comments and advice, and his translation of Euler's words in Sect. 2.2.2.

References

1. C. Altseimer, A.V. Borovik, *Probabilistic Recognition of Orthogonal and Symplectic Groups*, in Groups and Computation, III, vol. 8, Columbus, OH, 1999 (Ohio State University Mathematical Research Institute Publications/de Gruyter, Berlin, 2001), pp. 1–20
2. M. Aschbacher, On the maximal subgroups of the finite classical groups. Invent. Math. **76**(3), 469–514 (1984)
3. L. Babai, *Local Expansion of Vertex-Transitive Graphs and Random Generation in Finite Groups*, in 23rd ACM Symposium on Theory of Computing (ACM, New York, 1991), pp. 164–174
4. L. Babai, R. Beals, *A Polynomial-Time Theory of Black Box Groups. I*, in Groups St. Andrews 1997 in Bath, I. London Mathematical Society Lecture Note Series, vol. 260 (Cambridge University Press, Cambridge, 1999), pp. 30–64
5. L. Babai, E. Szemerédi, *On the Complexity of Matrix Group Problems I*, in 25th Annual Symposium on Foundations of Computer Science (IEEE Computer Society Press, Los Alamitos, 1984), pp. 229–240
6. L. Babai, W.M. Kantor, P.P. Pálfy, Á. Seress, Black-box recognition of finite simple groups of Lie type by statistics of element orders. J. Group Theor. **5**(4), 383–401 (2002)
7. L. Babai, R. Beals, Á. Seress, *Polynomial-Time Theory of Matrix Groups*, in 41st ACM Symposium on Theory of Computing, Bethesda, MD, 2009 (ACM, New York, 2009), pp. 55–64
8. R. Beals, C.R. Leedham-Green, A.C. Niemeyer, C.E. Praeger, Á. Seress, Permutations with restricted cycle structure and an algorithmic application. Combin. Probab. Comput. **11**(5), 447–464 (2002)
9. R. Beals, C.R. Leedham-Green, A.C. Niemeyer, C.E. Praeger, Á. Seress, A black-box group algorithm for recognizing finite symmetric and alternating groups. I. Trans. Am. Math. Soc. **355**(5), 2097–2113 (2003)
10. D.E.-C. Ben-Ezra, Counting elements in the symmetric group, Int. J. Algebra Comput. **19**(3), 305–313 (2009)
11. E.A. Bender, Asymptotic methods in enumeration. SIAM Rev. **16**, 485–515 (1974)

12. E.A. Bertram, B. Gordon, Counting special permutations. Eur. J. Comb. **10**(3), 221–226 (1989)
13. E.D. Bolker, A.M. Gleason, Counting permutations. J. Comb. Theor. Ser. A **29**(2), 236–242 (1980)
14. M. Bóna, A. McLennan, D. White, Permutations with roots. Random Struct. Algorithm **17**(2), 157–167 (2000)
15. W. Bosma, J. Cannon, C. Playoust, The Magma algebra system. I. The user language. J. Symbolic Comput. **24**, 235–265 (1997)
16. J.N. Bray, An improved method for generating the centralizer of an involution. Arch. Math. (Basel) **74**(4), 241–245 (2000)
17. R.W. Carter, *Finite Groups of Lie Type* (Wiley Classics Library, Wiley, Chichester, 1993), Conjugacy classes and complex characters, Reprint of the 1985 original, A Wiley-Interscience Publication
18. F. Celler, C.R. Leedham-Green, S.H. Murray, A.C. Niemeyer, E.A. O'Brien, Generating random elements of a finite group. Comm. Algebra **23**(13), 4931–4948 (1995)
19. W.W. Chernoff, Solutions to $x^r = \alpha$ in the alternating group. Ars Combin. **29**(C), 226–227 (1990) (Twelfth British Combinatorial Conference, Norwich, 1989)
20. S. Chowla, I.N. Herstein, W.K. Moore, On recursions connected with symmetric groups. I. Can. J. Math. **3**, 328–334 (1951)
21. S. Chowla, I.N. Herstein, W.R. Scott, The solutions of $x^d = 1$ in symmetric groups. Norske Vid. Selsk. Forh. Trondheim **25**, 29–31 (1952/1953)
22. A.M. Cohen, S.H. Murray, An algorithm for Lang's Theorem. J. Algebra **322**(3), 675–702 (2009)
23. A. de Moivre, *The Doctrine of Chances: Or, A Method of Calculating the Probability of Events in Play*, 2nd edn. (H. Woodfall, London, 1738)
24. J.D. Dixon, Generating random elements in finite groups. Electron. J. Comb. **15**(1), Research Paper 94 (2008)
25. P. Dusart, The kth prime is greater than $k(\ln k + \ln \ln k - 1)$ for $k \geq 2$. Math. Comp. **68**(225), 411–415 (2009)
26. P. Erdős, M. Szalay, *On Some Problems of the Statistical Theory of Partitions*, in Number Theory, vol. I, Budapest, 1987. Colloq. Math. Soc. János Bolyai, vol. 51 (North-Holland, Amsterdam, 1990), pp. 93–110
27. P. Erdős, P. Turán, On some problems of a statistical group-theory. I. Z. Wahrscheinlichkeitstheorie und Verw. Gebiete **4**, 175–186 (1965)
28. P. Erdős, P. Turán, On some problems of a statistical group-theory. II. Acta Math. Acad. Sci. Hung. **18**, 151–163 (1967)
29. P. Erdős, P. Turán, On some problems of a statistical group-theory. III. Acta Math. Acad. Sci. Hung. **18**, 309–320 (1967)
30. P. Erdős, P. Turán, On some problems of a statistical group-theory. IV. Acta Math. Acad. Sci. Hung. **19**, 413–435 (1968)
31. P. Erdős, P. Turán, On some problems of a statistical group theory. VI. J. Indian Math. Soc. **34**(3–4), 175–192 (1970/1971)
32. P. Erdős, P. Turán, On some problems of a statistical group theory. V. Period. Math. Hung. **1**(1), 5–13 (1971)
33. L. Euler, *Calcul de la probabilité dans le jeu de rencontre*. Mémoires de l'Academie des Sciences de Berlin, 7 (1751) 1753, pp. 255–270. Reprinted in Opera Omnia: Series 1, vol. 7, pp. 11–25. Available through The Euler Archive at www.EulerArchive.org.
34. L. Euler, *Solutio Quaestionis curiosae ex doctrina combinationum*. Mémoires de l'Académie des Sciences de St.-Petersbourg, 3:57–64, 1811. Reprinted in Opera Omnia: Series 1, vol. 7, pp. 435–440. Available through The Euler Archive at www.EulerArchive.org.
35. P. Flajolet, R. Sedgewick, *Analytic Combinatorics* (Cambridge University Press, Cambridge, 2009)
36. J. Fulman, P.M. Neumann, C.E. Praeger, A generating function approach to the enumeration of matrices in classical groups over finite fields. Mem. Am. Math. Soc. **176**(830), vi+90 (2005)

37. The GAP Group, GAP — Groups, Algorithms, and Programming, Version 4.5.2(beta), 2011, http://www.gap-system.org/
38. S.P. Glasby, Using recurrence relations to count certain elements in symmetric groups. Eur. J. Comb. **22**(4), 497–501 (2001)
39. W.M.Y. Goh, E. Schmutz, The expected order of a random permutation. Bull. Lond. Math. Soc. **23**(1), 34–42 (1991)
40. V. Gončarov, On the field of combinatory analysis. Am. Math. Soc. Transl. **19**(2), 1–46 (1962)
41. D. Gorenstein, R. Lyons, R. Solomon, *The Classification of the Finite Simple Groups*. Mathematical Surveys and Monographs, vol. 40 (American Mathematical Society, Providence, 1994)
42. O. Gruder, Zur Theorie der Zerlegung von Permutationen in Zyklen. Ark. Mat. **2**(5), 385–414 (1953)
43. W.K. Hayman, A generalisation of Stirling's formula. J. Reine Angew. Math. **196**, 67–95 (1956)
44. R.B. Herrera, The number of elements of given period in finite symmetric groups. Am. Math. Mon. **64**, 488–490 (1957)
45. P.E. Holmes, S.A. Linton, E.A. O'Brien, A.J.E. Ryba, R.A. Wilson, Constructive membership in black-box groups. J. Group Theor. **11**(6), 747–763 (2008)
46. I.M. Isaacs, W.M. Kantor, N. Spaltenstein, On the probability that a group element is p-singular. J. Algebra **176**(1), 139–181 (1995)
47. E. Jacobsthal, Sur le nombre d'éléments du groupe symétrique S_n dont l'ordre est un nombre premier. Norske Vid. Selsk. Forh. Trondheim **21**(12), 49–51 (1949)
48. W.M. Kantor, Á. Seress, Black box classical groups. Mem. Am. Math. Soc. **149**(708), viii+168 (2001)
49. A.V. Kolchin, Equations that contain an unknown permutation. Diskret. Mat. **6**(1), 100–115 (1994)
50. V.F. Kolchin, *Random Graphs*. Encyclopedia of Mathematics and Its Applications, vol. 53 (Cambridge University Press, Cambridge, 1999)
51. E. Landau, *Handbuch der Lehre von der Verteilung der Primzahlen. 2 Bände*, 2nd edn. (Chelsea Publishing Co., New York, 1953), With an appendix by Paul T. Bateman
52. C.R. Leedham-Green, *The Computational Matrix Group Project*, in Groups and Computation, III, vol. 8, Columbus, OH, 1999 (Ohio State University Mathematical Research Institute Publications/de Gruyter, Berlin, 2001), pp. 229–247
53. C.R. Leedham-Green, E.A. O'Brien, Constructive recognition of classical groups in odd characteristic. J. Algebra **322**(3), 833–881 (2009)
54. G.I. Lehrer, Rational tori, semisimple orbits and the topology of hyperplane complements. Comment. Math. Helv. **67**(2), 226–251 (1992)
55. G.I. Lehrer, The cohomology of the regular semisimple variety. J. Algebra **199**(2), 666–689 (1998)
56. M.W. Liebeck, E.A. O'Brien, Finding the characteristic of a group of Lie type. J. Lond. Math. Soc. (2) **75**(3), 741–754 (2007)
57. M.W. Liebeck, A. Shalev, The probability of generating a finite simple group. Geom. Dedicata **56**(1), 103–113 (1995)
58. F. Lübeck, A.C. Niemeyer, C.E. Praeger, Finding involutions in finite Lie type groups of odd characteristic. J. Algebra **321**(11), 3397–3417 (2009)
59. E.M. Luks, *Permutation Groups and Polynomial-Time Computation*, in Groups and computation, New Brunswick, NJ, 1991. DIMACS Series in Discrete Mathematics and Theoretical Computer Science, vol. 11 (American Mathematical Society, Providence, 1993), pp. 139–175
60. R. Lyons, Evidence for a new finite simple group. J. Algebra **20**, 540–569 (1972)
61. A. Maróti, Symmetric functions, generalized blocks, and permutations with restricted cycle structure. Eur. J. Comb. **28**(3), 942–963 (2007)
62. N. Metropolis, The beginnings of the Monte Carlo method. Los Alamos Sci. **15** (Special Issue), 125–130 (1987)

63. M.P. Mineev, A.I. Pavlov, The number of permutations of a special form. Mat. Sb. (N.S.) **99(141)**(3), 468–476, 480 (1976)
64. P.R. de Monmort, *Essay d'analyse sur les jeux de hazard* (J. Quillau, Paris, 1708)
65. P.R. de Monmort, *Essay d'analyse sur les jeux de hazard*, 2nd edn. (J. Quillau, Paris, 1713)
66. L. Moser, M. Wyman, On solutions of $x^d = 1$ in symmetric groups. Can. J. Math. **7**, 159–168 (1955)
67. L. Moser, M. Wyman, Asymptotic expansions. Can. J. Math. **8**, 225–233 (1956)
68. L. Moser, M. Wyman, Asymptotic expansions. II. Can. J. Math. **9**, 194–209 (1957)
69. P.M. Neumann, C.E. Praeger, A recognition algorithm for special linear groups. Proc. Lond. Math. Soc. (3) **65** (3), 555–603 (1992)
70. P.M. Neumann, C.E. Praeger, Cyclic matrices over finite fields. J. Lond. Math. Soc. (2) **52**, 263–284 (1995)
71. A.C. Niemeyer, T. Popiel, C.E. Praeger, Abundant p-singular elements in finite classical groups, preprint (2012) http://arxiv.org/abs/1205.1454v2
72. A.C. Niemeyer, C.E. Praeger, A recognition algorithm for classical groups over finite fields. Proc. Lond. Math. Soc. (3) **77** (1), 117–169 (1998)
73. A.C. Niemeyer, C.E. Praeger, On the frequency of permutations containing a long cycle. J. Algebra **300**(1), 289–304 (2006)
74. A.C. Niemeyer, C.E. Praeger, On permutations of order dividing a given integer. J. Algebr. Comb. **26**(1), 125–142 (2007)
75. A.C. Niemeyer, C.E. Praeger, On the proportion of permutations of order a multiple of the degree. J. Lond. Math. Soc. (2) **76**(3), 622–632 (2007)
76. A.C. Niemeyer, C.E. Praeger, Estimating proportions of elements in finite groups of Lie type. J. Algebra **324**(1), 122–145 (2010)
77. E.A. O'Brien, *Algorithms for Matrix Groups*, in Groups St. Andrews 2009 in Bath, vol. 2. London Mathematical Society Lecture Note Series, vol. 388 (Cambridge University Press, Cambridge, 2011), pp. 297–323
78. C.W. Parker, R.A. Wilson, Recognising simplicity of black-box groups by constructing involutions and their centralisers. J. Algebra **324**(5), 885–915 (2010)
79. E.T. Parker, P.J. Nikolai, A search for analogues of the Mathieu groups. Math. Tables Aids Comput. **12**, 38–43 (1958)
80. A.I. Pavlov, An equation in a symmetric semigroup. Trudy Mat. Inst. Steklov. **177**, 114–121, 208 (1986); Proc. Steklov Inst. Math. 1988(4), 121–129, Probabilistic problems of discrete mathematics
81. A.I. Pavlov, On permutations with cycle lengths from a fixed set. Theor. Probab. Appl. **31**, 618–619 (1986)
82. W. Plesken, D. Robertz, The average number of cycles. Arch. Math. (Basel) **93**(5), 445–449 (2009)
83. C.E. Praeger, On elements of prime order in primitive permutation groups. J. Algebra **60**(1), 126–157 (1979)
84. C.E. Praeger, Á. Seress, Probabilistic generation of finite classical groups in odd characteristic by involutions. J. Group Theor. **14**(4), 521–545 (2011)
85. C.E. Praeger, Á. Seress, Balanced involutions in the centralisers of involutions in finite general linear groups of odd characteristic (in preparation)
86. C.E. Praeger, Á. Seress, Regular semisimple elements and involutions in finite general linear groups of odd characteristic. Proc. Am. Math. Soc. **140**, 3003–3015 (2012)
87. Á. Seress, *Permutation Group Algorithms*. Cambridge Tracts in Mathematics, vol. 152 (Cambridge University Press, Cambridge, 2003)
88. C.C. Sims, *Computational Methods in the Study of Permutation Groups*, in Computational Problems in Abstract Algebra, Proceedings of the Conference, Oxford, 1967 (Pergamon, Oxford, 1970), pp. 169–183
89. C.C. Sims, *The Existence and Uniqueness of Lyons' Group*, in Finite groups '72, Proceedings of the Gainesville Conference, University of Florida, Gainesville, FL, 1972. North–Holland Mathematical Studies, vol. 7 (North-Holland, Amsterdam, 1973), pp. 138–141

90. A.N. Timashev, Random permutations with cycle lengths in a given finite set. Diskret. Mat.
     20(1), 25–37 (2008)
91. J. Touchard, Sur les cycles des substitutions. Acta Math. **70**(1), 243–297 (1939)
92. L.M. Volynets, The number of solutions of the equation $x^s = e$ in a symmetric group. Mat.
     Zametki **40**(2), 155–160, 286 (1986)
93. R. Warlimont, Über die Anzahl der Lösungen von $x^n = 1$ in der symmetrischen Gruppe S_n.
     Arch. Math. (Basel) **30** (6), 591–594 (1978)
94. H. Wielandt, *Finite Permutation Groups*, Translated from the German by R. Bercov (Academic,
     New York, 1964)
95. H.S. Wilf, The asymptotics of $e^{P(z)}$ and the number of elements of each order in S_n. Bull. Am.
     Math. Soc. (N.S.) **15** (2), 228–232 (1986)
96. H.S. Wilf, *Generatingfunctionology*, 2nd edn. (Academic, Boston, 1994)
97. K. Zsigmondy, Zur Theorie der Potenzreste. Monatsh. für Math. U. Phys. **3**, 265–284 (1892)

Chapter 3
Designs, Groups and Computing

Leonard H. Soicher

3.1 Introduction

In this chapter we present some applications of groups and computing to the discovery, construction, classification and analysis of combinatorial designs. The focus is on certain block designs and their statistical efficiency measures, and in particular semi-Latin squares, which are certain 1-designs with additional block structure and which generalise Latin squares.

The chapter starts with background material on block designs, statistical efficiency measures for 1-designs, permutation groups, Latin squares and semi-Latin squares. We then review statistical optimality results for semi-Latin squares. Next, we introduce the recent theory of "uniform" semi-Latin squares, which generalise complete sets of mutually orthogonal Latin squares, and are statistically "Schur-optimal". We then describe a recent construction which determines a semi-Latin square SLS(G) from a transitive permutation group G, the square being uniform precisely when G is 2-transitive. Moreover, we show how certain structural properties of SLS(G) are determined from the structure of G, and (due to Martin Liebeck) how statistical efficiency measures of SLS(G) are determined from the degrees and multiplicities of the irreducible constituents of the permutation character of G.

We then turn to computation, and discuss the DESIGN package [34] for GAP [18], focussing on the function BlockDesigns, which can be used for a wide variety of block design classifications using groups, and the function BlockDesignEfficiency, for exact information on the efficiency measures of a given 1-design. We then show how the BlockDesigns function can be applied to obtain classifications of semi-Latin squares, and of uniform semi-Latin squares and their subsquares, which can then be analysed using the BlockDesign-

L.H. Soicher (✉)
School of Mathematical Sciences, Queen Mary University of London, Mile End Road, London E1 4NS, UK
e-mail: L.H.Soicher@qmul.ac.uk

A. Detinko et al. (eds.), *Probabilistic Group Theory, Combinatorics, and Computing*, Lecture Notes in Mathematics 2070, DOI 10.1007/978-1-4471-4814-2_3,
© Springer-Verlag London 2013

Efficiency function. We give an extended example of this, including the determination of the first published efficient $(6 \times 6)/k$ semi-Latin squares, for $k = 7, 8, 9$. It is hoped that our examples of the use of the DESIGN package will help the reader to use this package in their own investigations of designs. We conclude the chapter with some open problems.

3.2 Background Material

3.2.1 Block Designs

A *block design* is an ordered pair $(V, \mathcal{B})$, such that V is a finite non-empty set of *points*, and $\mathcal{B}$ is a (disjoint from V) finite multiset (or collection) of non-empty subsets of V called *blocks*, such that every point is in at least one block.

In a multiset (of blocks say), order does not matter, but the number of times an element occurs (its *multiplicity*) does indeed matter. We denote a multiset with elements $A_1, \ldots, A_b$ (including any repeated elements) by $[A_1, \ldots, A_b]$. For example, the block design $(V, \mathcal{B})$, with $V = \{1, 2, 3\}$ and

$$\mathcal{B} = [\{1, 2\}, \{1, 2, 3\}, \{1, 2, 3\}, \{1, 3\}, \{2, 3\}],$$

has three points and five blocks.

Let t be a non-negative integer. A t-*design*, or more specifically a t-(v, k, λ) *design*, is a block design $(V, \mathcal{B})$ such that $v = |V|$, each block has the same size k, and each t-subset of V is contained in the same positive number λ of blocks. It is well known that a t-design is also an s-design, for $s = 0, \ldots, t-1$ (see, for example, [23, Theorem 19.3]). For example, the block design $(V, \mathcal{B})$, with $V = \{1, \ldots, 7\}$ and

$$\mathcal{B} = [\{1, 2, 3\}, \{1, 4, 5\}, \{1, 6, 7\}, \{2, 4, 7\}, \{2, 5, 6\}, \{3, 4, 6\}, \{3, 5, 7\}],$$

is a 2-$(7, 3, 1)$ design, and also a 1-$(7, 3, 3)$ design and a 0-$(7, 3, 7)$ design.

The dual of a block design is obtained by interchanging the roles of points and blocks. More precisely, if $\Delta = (\{\alpha_1, \ldots, \alpha_v\}, [A_1, \ldots, A_b])$ is a block design, then the *dual* Δ^* of Δ is the block design $(\{1, \ldots, b\}, [B_1, \ldots, B_v])$, with $B_i = \{j : \alpha_i \in A_j\}$. Note that if Δ is a 1-(v, k, r) design then Δ^* is a 1-$(vr/k, r, k)$ design.

Two block designs $\Delta_1 = (V_1, \mathcal{B}_1)$ and $\Delta_2 = (V_2, \mathcal{B}_2)$ are *isomorphic* if there is a bijection from V_1 to V_2 that maps $\mathcal{B}_1$ to $\mathcal{B}_2$. The set of all isomorphisms from a block design Δ to itself forms a group, the *automorphism group* Aut(Δ) of Δ.

Block designs are of interest to both pure mathematicians and statisticians. They are used by statisticians for the design of comparative experiments: the points represent "treatments" to be compared, and the blocks represent homogeneous testing material, so usually only within-blocks information is used in the comparison of treatment effects. Although statisticians sometimes allow points to be repeated within a block, here we do not.

3.2.2 Efficiency Measures of 1-Designs

We now discuss certain statistical efficiency measures of 1-designs. The basic idea is that once you have decided that your experimental design needs to be in a certain class $\mathscr{C}$ of 1-(v, k, r) designs, you want to choose a design in $\mathscr{C}$ able to give as much information as possible; that is, the most "efficient" design in $\mathscr{C}$ with respect to one or more of the efficiency measures defined below. The reader who wants to learn more about statistical design theory and the theory of optimal designs should consult the excellent survey article [6], which was written for combinatorialists. Other useful references for these topics include [3, 4, 11, 29].

Let Δ be a 1-(v, k, r) design, with $v \geq 2$. The *concurrence matrix* of Δ is the $v \times v$ matrix whose rows and columns are indexed by the points, and whose (α, β)-entry is the number of blocks containing both points α and β. The *scaled information matrix* of Δ is

$$F(\Delta) := I_v - (rk)^{-1}\Lambda,$$

where I_v is the $v \times v$ identity matrix and Λ is the concurrence matrix of Δ. The matrix $F(\Delta)$ is real, symmetric and positive semi-definite, and is scaled so that its (all real) eigenvalues lie in the interval $[0, 1]$. Moreover, $F(\Delta)$ has constant row-sum 0, so the all-1 vector is an eigenvector with corresponding eigenvalue 0. It can be shown that the remaining eigenvalues are all non-zero if and only if Δ is connected (i.e. its point-block incidence graph is connected). Omitting the zero eigenvalue corresponding to the all-1 vector, the eigenvalues

$$\delta_1 \leq \delta_2 \leq \cdots \leq \delta_{v-1}$$

of $F(\Delta)$ are called the *canonical efficiency factors* of Δ. Note that the canonical efficiency factors of two isomorphic 1-designs are the same.

If Δ is not connected, then we define $A_\Delta = D_\Delta = E_\Delta := 0$. Otherwise, we define these *efficiency measures* by

$$A_\Delta := (v-1) / \sum_{i=1}^{v-1} 1/\delta_i,$$

$$D_\Delta := \left(\prod_{i=1}^{v-1} \delta_i \right)^{1/(v-1)},$$

$$E_\Delta := \delta_1 = \min\{\delta_1, \ldots, \delta_{v-1}\}.$$

Note that A_Δ (respectively D_Δ) is the harmonic mean (respectively geometric mean) of the canonical efficiency factors of Δ.

If $k = v$ (so each block contains every point) then each canonical efficiency factor of Δ is equal to 1 and so is each of the efficiency measures above. If $k < v$, we want to minimize the loss of "information" due to being forced to use "incomplete" blocks, and want the above efficiency measures to be as close to 1 as possible.

We now define optimality in a class of 1-designs with respect to a given efficiency measure. The 1-(v, k, r) design Δ is *A-optimal* in a class $\mathscr{C}$ of 1-(v, k, r) designs containing Δ if $A_\Delta \geq A_\Gamma$ for each $\Gamma \in \mathscr{C}$. D-optimal and E-optimal are defined similarly. We say Δ is *Schur-optimal* in a class $\mathscr{C}$ of 1-(v, k, r) designs containing Δ if for each design $\Gamma \in \mathscr{C}$, with canonical efficiency factors $\gamma_1 \leq \cdots \leq \gamma_{v-1}$, we have

$$\sum_{i=1}^{\ell} \delta_i \geq \sum_{i=1}^{\ell} \gamma_i,$$

for $\ell = 1, \ldots, v - 1$. A Schur-optimal design need not exist within a given class $\mathscr{C}$ of 1-(v, k, r) designs, but when it does, that design is optimal in $\mathscr{C}$ with respect to a very wide range of statistical optimality criteria, including being A-, D- and E-optimal [19].

It is not difficult to see that if Δ is a 2-(v, k, λ) design, then the canonical efficiency factors of Δ are all equal (to $v(k - 1)/((v - 1)k)$), from which it follows that Δ is Schur-optimal in the class of all 1-$(v, k, \lambda(v - 1)/(k - 1))$ designs. However, a 2-design may well not exist with the properties we are interested in.

The canonical efficiency factors of a 1-design Δ not equal to 1, and their multiplicities, are the same as those of the dual block design Δ^* of Δ (see [6, Sect. 3.1.1]). We thus obtain the following:

Theorem 3.1. *A* 1-(v, k, r) *design* Δ *is A-optimal (respectively D-optimal, E-optimal, Schur-optimal) in a class* $\mathscr{C}$ *of* 1-(v, k, r) *designs if and only if* Δ^* *is A-optimal (respectively D-optimal, E-optimal, Schur-optimal) in the class consisting of the dual block designs of the designs in* $\mathscr{C}$.

3.2.3 Permutation Groups

We now review some basic definitions and results in the theory of permutation groups and group actions. An excellent reference for permutation groups is [10].

A *permutation group* G on a set V of *points* is a subgroup of the group $\mathrm{Sym}(V)$ of all permutations of V, with the group operation being composition. The image of $\alpha \in V$ under $g \in G$ is denoted αg (our permutations act on the right). The *degree* of G is the size of V. The *symmetric group of degree* n, denoted S_n, is the group $\mathrm{Sym}(\{1, \ldots, n\})$ of all permutations of $\{1, \ldots, n\}$.

An *action* of a group G on a set V is a function $\psi : V \times G \to V$, with $(\alpha, g)\psi$ denoted α^g, such that $\alpha^1 = \alpha$ and $(\alpha^g)^h = \alpha^{gh}$, for all $\alpha \in V$ and all $g, h \in G$. Given an action of G on V, we say that G *acts on* V.

If G is a permutation group on V then G acts *naturally* on V, where $\alpha^g := \alpha g$. An action of G on V gives rise to a homomorphism $\phi : G \to \mathrm{Sym}(V)$ defined by $\alpha(g\phi) := \alpha^g$ for all $\alpha \in V$ and $g \in G$.

Suppose G acts on V and $\alpha \in V$. The G-*orbit* of α is $\alpha^G := \{\alpha^g : g \in G\}$. The set $\{\alpha^G : \alpha \in V\}$ of all G-orbits is a partition of V. The *stabilizer* in G of α is $G_\alpha := \{g \in G : \alpha^g = \alpha\}$. This stabilizer G_α is a subgroup of G, and if G is finite then $|G| = |G_\alpha||\alpha^G|$.

Let G act on a finite set V. Then G has actions on many sets. Here are some:

- G acts on the set of all t-tuples of distinct elements of V, where $(\alpha_1, \ldots, \alpha_t)^g := (\alpha_1^g, \ldots, \alpha_t^g)$.
- G acts on the set of all subsets of V, where $\{\alpha_1, \ldots, \alpha_k\}^g := \{\alpha_1^g, \ldots, \alpha_k^g\}$.
- G acts on the set of all finite multisets of subsets of V, where $[A_1, \ldots, A_b]^g := [(A_1)^g, \ldots, (A_b)^g]$.
- G acts on the set of all block designs with point set V, where $(V, \mathscr{B})^g := (V, \mathscr{B}^g)$.

A permutation group G on a non-empty set V is *transitive* if for every $\alpha, \beta \in V$ there is a $g \in G$ with $\alpha g = \beta$ (i.e. there is just one G-orbit in the natural action of G on V). More generally, a permutation group G on a set V of size at least t is t-*transitive* if for every pair $(\alpha_1, \ldots, \alpha_t)$, $(\beta_1, \ldots, \beta_t)$ of t-tuples of distinct elements of V, there is a $g \in G$ with $(\alpha_1 g, \ldots, \alpha_t g) = (\beta_1, \ldots, \beta_t)$. It is easy to see that if G is a t-transitive permutation group on a finite set V of size $v \geq t$, and B is a subset of V of size $k \geq t$, then the block design with point set V and block (multi)set the G-orbit B^G is a t-(v, k, λ) design, with $\lambda = |B^G|\binom{k}{t}/\binom{v}{t}$.

3.2.4 Latin Squares

A *Latin square* of *order n* is an $n \times n$ array L, whose entries are elements of an n-set Ω, the set of *symbols* for L, such that each symbol occurs exactly once in each row and exactly once in each column of L. For example, a completed Sudoku puzzle is a special kind of Latin square of order 9.

Two Latin squares L_1 and L_2 of order n, with respective symbol sets Ω_1 and Ω_2, are *orthogonal* if for every $\alpha_1 \in \Omega_1$ and $\alpha_2 \in \Omega_2$, there is an (i, j) such that α_1 is the (i, j)-entry in L_1 and α_2 is the (i, j)-entry in L_2. For example, here are two orthogonal Latin squares of order 3:

1	2	3		4	5	6
3	1	2		5	6	4
2	3	1		6	4	5

Latin squares $L_1, \ldots, L_m$ of order n are said to be *mutually orthogonal* if they are pairwise orthogonal, in which case $\{L_1, \ldots, L_m\}$ is called a set of *mutually orthogonal Latin squares* or a set of *MOLS*.

Let $n > 1$. A set of MOLS of order n has size at most $n - 1$, and the existence of a set of MOLS of order n having size $n - 1$ (called a *complete set* of MOLS) is equivalent to the existence of a projective plane of order n. A complete set of MOLS of order n exists when n is a prime power, but it is a famous problem as to whether a complete set of MOLS of order n exists for some non-prime-power n.

3.2.5 Semi-Latin Squares

An $(n \times n)/k$ *semi-Latin square* is an $n \times n$ array S, whose entries are k-subsets of an nk-set Ω, the set of *symbols* for S, such that each symbol is in exactly one entry in each row and exactly one entry in each column of S. The entry in row i and column j is called the (i, j)-entry of S and is denoted by $S(i, j)$. To avoid trivialities, we assume throughout that $n > 1$, $k > 0$. Note that an $(n \times n)/1$ semi-Latin square is (essentially) the same thing as a Latin square of order n. We consider two $(n \times n)/k$ semi-Latin squares to be *isomorphic* if one can be obtained from the other by applying one or more of: a row permutation, a column permutation, transposing, and renaming symbols.

Semi-Latin squares have many applications, including the design of agricultural experiments, consumer testing, and message authentication (see [1, 2, 5, 17, 26]). Semi-Latin squares exist in profusion, and a good choice of semi-Latin square for a given application can be very important.

Let s be a positive integer. An *s-fold inflation* of an $(n \times n)/k$ semi-Latin square is obtained by replacing each symbol α in the semi-Latin square by s symbols $\sigma_{\alpha,1}, \ldots, \sigma_{\alpha,s}$, such that $\sigma_{\alpha,i} = \sigma_{\beta,j}$ if and only if $\alpha = \beta$ and $i = j$. The result is an $(n \times n)/(ks)$ semi-Latin square. For example, here is a $(3 \times 3)/2$ semi-Latin square formed by a twofold inflation of a Latin square of order 3:

1 4	2 5	3 6
3 6	1 4	2 5
2 5	3 6	1 4

The *superposition* of an $(n \times n)/k$ semi-Latin square with an $(n \times n)/\ell$ semi-Latin square (with disjoint symbol sets) is performed by superimposing the first square upon the second, resulting in an $(n \times n)/(k + \ell)$ semi-Latin square. For example, here is a $(3 \times 3)/2$ semi-Latin square which is the superposition of two (mutually orthogonal) Latin squares of order 3:

$$
\begin{array}{|c|c|c|}
\hline
1\,4 & 2\,5 & 3\,6 \\
\hline
3\,5 & 1\,6 & 2\,4 \\
\hline
2\,6 & 3\,4 & 1\,5 \\
\hline
\end{array}
\tag{3.1}
$$

An $(n \times n)/k$ semi-Latin square in which any two distinct symbols occur together in at most one block is called a SOMA(k, n) (SOMA is an acronym for "simple orthogonal multi-array" [25]). For example, the semi-Latin square (3.1) is a SOMA$(2, 3)$, and more generally, the superposition of k MOLS of order n is a SOMA(k, n). However, a SOMA(k, n) need not be a superposition of MOLS, and may exist even when there do not exist k MOLS of order n. However, it is easy to see that if a SOMA(k, n) exists then $k < n$.

The *dual* S^* of an $(n \times n)/k$ semi-Latin square S is a block design with point set $\{1, \ldots, n\}^2$ and nk blocks, one for each symbol of S, with the block for a symbol α consisting precisely of the ordered pairs (i, j) such that $\alpha \in S(i, j)$. Each block of S^* is of the form $[(1, 1^g), \ldots, (n, n^g)]$ for some $g \in S_n$, and S^* is a 1-(n^2, n, k) design, which may have repeated blocks. Up to the naming of its symbols, the semi-Latin square S can be recovered from S^*, so S^* really represents the class of semi-Latin squares obtainable from S by renaming symbols. For example, let S be the semi-Latin square (3.1). Then $S^* = (V, \mathscr{B})$, with $V = \{1, 2, 3\}^2$ and

$$\mathscr{B} = \begin{array}{l} [\{(1,1),(2,2),(3,3)\}, \ \{(1,2),(2,3),(3,1)\}, \\ \{(1,3),(2,1),(3,2)\}, \ \{(1,1),(2,3),(3,2)\}, \\ \{(1,2),(2,1),(3,3)\}, \ \{(1,3),(2,2),(3,1)\}]. \end{array}$$

3.3 Optimality Results for Semi-Latin Squares

Let S be an $(n \times n)/k$ semi-Latin square. If we ignore the row and column structure of S, we obtain its *underlying block design* $\Delta(S)$, whose points are the symbols of S and whose blocks are the entries of S. Note that $\Delta(S)$ is a 1-(nk, k, n) design, and that the dual S^* of S is isomorphic as a block design to the dual of $\Delta(S)$. However, S^* has a structured point-set, unlike $\Delta(S)^*$.

Following the analysis of Bailey [2], S is *optimal* with respect to a given statistical optimality criterion if and only if $\Delta(S)$ is optimal with respect to that criterion in the class of underlying block designs of $(n \times n)/k$ semi-Latin squares. For this reason, we say that S has canonical efficiency factors, or a given efficiency measure, when $\Delta(S)$ has those canonical efficiency factors, or that efficiency measure.

Various optimality results for $(n \times n)/k$ semi-Latin squares are known. These include:

- Cheng and Bailey [13] proved that a superposition of k MOLS of order n (with pairwise disjoint symbol sets) is A-, D- and E-optimal.
- Bailey [2] proved that for all $s \geq 1$, an s-fold inflation of a superposition of $n - 1$ MOLS of order n is A-, D- and E-optimal.
- Bailey [2] classified the 3×3 semi-Latin squares and determined the ones that are A-, D- and E-optimal.

- Chigbu [14] determined the $(4 \times 4)/4$ semi-Latin squares that are A-, D- and E-optimal.
- Using the computational methods described in this chapter, Soicher [37] classified the $(4 \times 4)/k$ semi-Latin squares for $k = 5, \ldots, 10$, and determined those that are A-, D- and E-optimal.

It is widely believed that when a SOMA(k, n) exists, then one that is optimal in the class of all SOMA(k, n)s is in fact optimal in the class of all $(n \times n)/k$ semi-Latin squares. There are not even two MOLS of order 6, but there are SOMA$(2, 6)$s and SOMA$(3, 6)$s. These have been classified and the best with respect to various measures of efficiency have been determined (see [5, 8, 25, 27, 31, 37]). There is no SOMA$(k, 6)$ with $k > 3$. It is not known whether there are three MOLS of order 10, but there are SOMA$(3, 10)$s and SOMA$(4, 10)$s (see [5, 30, 31]).

3.4   Uniform Semi-Latin Squares

We now introduce the concept of uniform semi-Latin squares, and some results about such squares from [36]. Uniform semi-Latin squares provide Schur-optimal semi-Latin squares of many sizes for which no optimal semi-Latin square was previously known for any optimality criterion.

An $(n \times n)/k$ semi-Latin square S is *uniform* if every two entries of S, not in the same row or column, intersect in a constant number $\mu = \mu(S)$ of points. For example, the semi-Latin square (3.1) is uniform, with $\mu = 1$.

Note that if S is a uniform semi-Latin square then an s-fold inflation of S is also uniform, and if S and T are both $n \times n$ uniform semi-Latin squares (with disjoint symbol sets) then the superposition of S and T is also uniform.

Lemma 3.2 ([36]). *If S is a uniform $(n \times n)/k$ semi-Latin square then*

$$\mu(S) = k/(n - 1),$$

and in particular, $n - 1$ divides k.

Theorem 3.3 ([36]). *An $(n \times n)/(n - 1)$ semi-Latin square S is uniform if and only if S is a superposition of $n - 1$ MOLS of order n.*

Uniform semi-Latin squares can be seen as generalising the concept of complete sets of MOLS. Since the μ-fold inflation of a uniform semi-Latin square is uniform, we see that the existence of a uniform $(n \times n)/((n - 1)\mu)$ semi-Latin square for all integers $\mu > 0$ is equivalent to the existence of a complete set of MOLS of order n, and such a complete set exists when n is a prime power. Although there are not even two MOLS of order 6, the following is proved in [36].

Theorem 3.4 ([36]). *There exist uniform $(6 \times 6)/(5\mu)$ semi-Latin squares for all integers $\mu > 1$.*

The statistical importance of uniform semi-Latin squares is due to the following:

Theorem 3.5 ([36]). *Suppose that S is a uniform $(n \times n)/k$ semi-Latin square. Then S is Schur-optimal; that is, $\Delta(S)$ is Schur-optimal in the class of underlying block designs of $(n \times n)/k$ semi-Latin squares.*

Proof. We give an outline of the proof. Details can be found in [36].

- The dual S^* of S is a partially balanced incomplete-block design with respect to the L_2-type association scheme (see [3]), so we may easily determine the eigenvalues of the concurrence matrix of S^* (see [39]), and hence the canonical efficiency factors of S^*.
- These canonical efficiency factors are $1 - 1/(n - 1)$, with multiplicity $(n - 1)^2$, and 1, with multiplicity $2(n - 1)$.
- The dual T^* of any $(n \times n)/k$ semi-Latin square T has at most $(n-1)^2$ canonical efficiency factors not equal to 1.
- It follows that S^* is Schur-optimal in the class of duals of $(n \times n)/k$ semi-Latin squares, and so $\Delta(S)$ is Schur-optimal in the class of underlying block designs of $(n \times n)/k$ semi-Latin squares. □

3.5 Semi-Latin Squares from Transitive Permutation Groups

In [36], a simple construction is given which produces a semi-Latin square SLS(G) from a transitive permutation group G, and in this section we discuss how properties of G determine properties of SLS(G).

Let G be a transitive permutation group on $\{1, \ldots, n\}$, with $n > 1$. For all $i, j \in \{1, \ldots, n\}$, there are exactly $|G|/n$ elements of G mapping i to j (the elements mapping i to j are precisely those in the right coset $G_i g$, where G_i is the stabilizer in G of i and g is any element of G with $ig = j$). Thus G defines a semi-Latin square, as follows. The $(n \times n)/k$ semi-Latin square SLS(G), with $k := |G|/n$, has symbol set G itself, and the symbol g is in the (i, j)-entry of SLS(G) if and only if $ig = j$. For example,

$$\text{SLS}(S_3) = \begin{array}{|cc|cc|cc|}
\hline
1 & (23) & (12) & (123) & (13) & (132) \\
\hline
(12) & (132) & 1 & (13) & (23) & (123) \\
\hline
(13) & (123) & (23) & (132) & 1 & (12) \\
\hline
\end{array}.$$

Theorem 3.6 ([36]). *Let G be a transitive permutation group on $\{1, \ldots, n\}$, with $n > 1$, and let $S := \text{SLS}(G)$. Then:*

- *Let H be a transitive subgroup of G of index m. Then S is a superposition of m semi-Latin squares, each isomorphic to SLS(H) (this comes from the partition of*

*G into the right (or left) cosets of H). In particular, if H has order n then S is a
superposition of $|G|/n$ isomorphic Latin squares.*

- *G contains a non-identity element with exactly f fixed points if and only if there
are two distinct symbols of S which occur together in exactly f entries of S.*
- *G is a Frobenius group (that is, a transitive permutation group in which no non-
identity element fixes more than one point) if and only if S is a superposition of
MOLS.*
- *$\Delta(S)$ is connected if and only if G has no normal subgroup N satisfying $G_1 \le
N \ne G$.*
- *The automorphism group of S (i.e. the group of all isomorphisms from S to S)
has structure*

$$(G \times G).((N_{S_n}(G)/G) \times C_2)$$

(in ATLAS [15] notation).
- *G is 2-transitive if and only if S is uniform.*

Proof. See [36] for the proofs of these assertions. We only repeat the (easy) proof
of the important last statement. Also see [38].

Suppose G is 2-transitive. Then for every $i, i', j, j' \in \{1, \ldots, n\}$ with $i \ne i'$ and
$j \ne j'$, there are precisely $\mu := |G|/(n(n-1))$ elements $g \in G$ with $ig = j$ and
$i'g = j'$. Thus, $S(i, j)$ and $S(i', j')$ intersect in exactly these μ elements, and so
S is uniform.

Conversely, suppose S is uniform. Then if $i, i', j, j' \in \{1, \ldots, n\}$ with $i \ne i'$
and $j \ne j'$, then $S(i, j)$ and $S(i', j')$ intersect in $\mu := k/(n-1) > 0$ symbols
(recall that $n > 1$, $k > 0$), so there is an element of G mapping i to j and i' to j'.
Thus G is 2-transitive.   □

Using the Classification of Finite Simple Groups, all the finite 2-transitive
permutation groups have been classified, and tables of these groups are given in
[10]. Each 2-transitive group G gives rise to a uniform semi-Latin square SLS(G),
certain properties of which can be deduced from properties of G. For example,
consideration of the groups $PGL_2(q)$ and $PSL_2(q)$, of degree $q + 1$ where q is a
prime power, yields the following result.

Theorem 3.7 ([36]). *Let q be a prime power. Then there exists a uniform, and
hence Schur-optimal, $((q + 1) \times (q + 1))/(q(q - 1))$ semi-Latin square S which
is a superposition of isomorphic Latin squares and in which every two distinct
symbols occur together in at most two entries. Moreover, if q is odd then S is also a
superposition of two isomorphic uniform $((q+1) \times (q+1))/(q(q-1)/2)$ semi-Latin
squares.*

3.5.1 The Canonical Efficiency Factors of SLS(G)

Let G be a transitive permutation group on $\{1, \ldots, n\}$, with $n > 1$, and let Λ be
the concurrence matrix of the underlying block design of SLS(G). Then Λ is a

$|G| \times |G|$ matrix whose rows and columns are indexed by the elements of G and whose (g, h)-entry is the number of fixed points of $g^{-1}h$, which is $\pi(g^{-1}h)$, where π is the permutation character of G. Applying this observation, Martin Liebeck (at the Fifth de Brún Workshop itself) discovered and proved the theorem below. The statement of the theorem and its proof use basic representation theory of finite groups over the complex numbers, such as can be found in [20].

Theorem 3.8 (M. Liebeck). *Let G be a transitive permutation group of degree $n > 1$ with permutation character π. Then the canonical efficiency factors of (the underlying block design of) SLS(G) are*

$$1 - \langle \chi, \pi \rangle / \chi(1),$$

repeated $\chi(1)^2$ times, where χ runs over the non-trivial complex irreducible characters of G.

Proof. Suppose $|G| = nk$, and let $GL_{nk}(\mathbb{C})$ be the group of all invertible $nk \times nk$ matrices over the complex numbers, whose rows and columns are indexed by the elements of G, and let $\rho : G \to GL_{nk}(\mathbb{C})$ be defined by $\rho(x)_{g,h} = 1$ if $gx = h$ and $\rho(x)_{g,h} = 0$ otherwise. In other words, ρ is the right-regular matrix representation of G, with natural G-module $V := \mathbb{C}^{nk}$.

Now let $C_0 = \{1\}, C_1, \ldots, C_d$ be the conjugacy classes of G, with respective representatives $c_0, \ldots, c_d$, let $\chi_0, \ldots, \chi_d$ be the complex irreducible characters of G, with χ_0 the trivial character, and let

$$A_i := \sum_{c \in C_i} \rho(c).$$

Then the matrices A_i commute pairwise, and V decomposes into a direct sum of common eigenspaces $V_0, \ldots, V_d$ of $A_0, \ldots, A_d$, with V_j being a G-submodule of V isomorphic to the direct sum of $\chi_j(1)$ copies of the irreducible G-module with character χ_j. In particular V_j has dimension $\chi_j(1)^2$, and V_j is an eigenspace for A_i with corresponding eigenvalue

$$|C_i| \chi_j(c_i) / \chi_j(1).$$

Note that V_0 is spanned by the all-1 vector.

Now A_i is a $(0, 1)$-matrix, with (g, h)-entry equal to 1 if and only if $g^{-1}h \in C_i$, and so

$$\Lambda = \sum_{i=0}^{d} \pi(c_i) A_i.$$

Thus, for $j = 0, \ldots, d$, Λ has eigenvalue

$$\sum_{i=0}^{d} \pi(c_i)|C_i|\chi_j(c_i)/\chi_j(1) = \sum_{g \in G} \chi_j(g)\pi(g^{-1})/\chi_j(1) = |G|\langle \chi_j, \pi \rangle/\chi_j(1),$$

repeated $\chi_j(1)^2$ times, where $\langle \, , \, \rangle$ denotes the standard inner product of characters (so $\langle \chi_j, \pi \rangle$ is the multiplicity of the irreducible character χ_j as a constituent of the permutation character π).

The scaled information matrix of the underlying block design Δ of SLS(G) is

$$F(\Delta) := I_{nk} - (nk)^{-1}\Lambda,$$

and so the canonical efficiency factors of SLS(G) are

$$1 - \langle \chi_j, \pi \rangle/\chi_j(1),$$

repeated $\chi_j(1)^2$ times, for $j = 1, \ldots, d$.   □

Thus, the canonical efficiency factors of SLS(G), and its A-, D- and E-efficiency measures can be determined from the degrees and multiplicities of the irreducible constituents of the permutation character of G. This has been done by Eamonn O'Brien (using MAGMA [9]) for all transitive permutation groups of degree ≤ 23, and by Soicher (using GAP) for the primitive permutation groups of non-prime-power degree $n \leq 500$ and with order $\leq n(n-1)$.

As an illustrative example, let G be the group A_5 in its primitive permutation representation of degree 10. The permutation character of G (in ATLAS [15] notation) decomposes as $1a + 4a + 5a$, so the canonical efficiency factors of SLS(G) are 3/4 (with multiplicity 16), 4/5 (with multiplicity 25), and 1 (with multiplicity 18). If six MOLS of order 10 exist, then their superposition S would have canonical efficiency factors 5/6 (with multiplicity 54) and 1 (with multiplicity 5); see [2, Corollary 5.2]. The ratios of the A-, D- and E-efficiency measures of SLS(G) with those of S are respectively approximately 0.9889, 0.9943 and 0.9, so in the (likely) absence of six MOLS of order 10, SLS(G) provides a highly efficient (and possibly optimal) $(10 \times 10)/6$ semi-Latin square.

3.6   The DESIGN Package

GAP [18] is an internationally developed, freely available, Open Source system for Computational Group Theory and related areas in algebra and combinatorics. The DESIGN package [34] is a refereed and officially accepted GAP package which provides functionality for constructing, classifying, partitioning and analysing block designs.

In this section we focus on the DESIGN package functions BlockDesigns, used to classify block designs, and BlockDesignEfficiency, used to determine efficiency measures of 1-designs. There are many other functions in the DESIGN package to construct and analyse block designs, including isomorphism testing and automorphism group computation, and functions to determine information about t-designs from their parameters. The best reference for all this, and for precise details on the parameters for the functions discussed here, is the DESIGN package documentation, which includes many examples. The details of the techniques used in the DESIGN package can be found in its documented Open Source code.

3.6.1   The BlockDesigns Function

The most important DESIGN package function is BlockDesigns, which can construct and classify block designs satisfying a wide range of user-specified properties. The properties which must be specified are:

- The number v of points (the point set is then $\{1, \ldots, v\}$, although the points may also be given names).
- The possible block sizes.
- For a given t, for each t-subset T of the points, the number of blocks containing T (this number may depend on T).

The properties which may optionally be additionally specified are:

- The maximum multiplicity of a block, for each possible block-size.
- The total number b of blocks.
- The block-size distribution.
- A replication number r (that is, specifying that every point is in exactly r blocks).
- The possible sizes of intersections of pairs of blocks of given sizes.
- A subgroup G of S_v, to specify that G-orbits of block designs are isomorphism classes (default: $G = S_v$, giving the usual notion of isomorphism).
- A subgroup H of G, such that H is required to be a subgroup of the automorphism group of each returned design (default: $H = \{1\}$, but specifying a non-trivial H can be a very powerful constraint; see [16, 21, 32, 33]).
- Whether the user wants a single design with the specified properties (if one exists), a list of G-orbit representatives of all such designs (i.e. isomorphism class representatives as determined by G; this is the default), or a list of distinct such designs containing at least one representative from each G-orbit.

The BlockDesigns function works by transforming the design classification problem into a problem of classifying cliques with a given vertex-weight sum in a certain graph whose vertices are "weighted" with non-zero vectors of non-negative integers. Each vertex of this graph represents a possible H-orbit of blocks, each with the same specified multiplicity, with two distinct vertices not joined by an edge only when the totality of the blocks they represent cannot be a submultiset of the blocks of a required design. Such a non-edge may be a result of user-specified properties of the required designs, or may be determined by applying

block intersection polynomials [12, 33]. The graph problem is then handled by
the GRAPE [35] function CompleteSubgraphsOfGivenSize, which uses
a complicated backtrack search. The reader may wish to consult the reference [21],
which gives detailed information on techniques used to classify block designs.

We now give some straightforward examples of the use of the BlockDesigns
function. A more complicated example will follow in Sect. 3.7. Note that first we
load the DESIGN package, which also loads the GRAPE package for graphs and
groups, which is heavily used by the DESIGN package.

```
gap> LoadPackage("design");
-----------------------------------------------------------------
Loading  GRAPE 4.5 (GRaph Algorithms using PErmutation groups)
by Leonard H. Soicher (http://www.maths.qmul.ac.uk/~leonard/).
Homepage: http://www.maths.qmul.ac.uk/~leonard/grape/
-----------------------------------------------------------------
-----------------------------------------------------------------
Loading  DESIGN 1.6 (The Design Package for GAP)
by Leonard H. Soicher (http://www.maths.qmul.ac.uk/~leonard/).
Homepage: http://www.designtheory.org/software/gap_design/
-----------------------------------------------------------------
true
```

We now classify the 2-(7, 3, 1) designs.

```
gap> designs:=BlockDesigns(rec( v:=7, blockSizes:=[3],
>      tSubsetStructure:=rec(t:=2, lambdas:=[1] ) ) );
[ rec( autGroup := Group([ (1,2)(5,7), (1,2,3)(5,7,6),
        (1,2,3)(4,7,5), (1,5,3)(2,4,7) ]),
   blockNumbers := [ 7 ], blockSizes := [ 3 ],
   blocks := [ [ 1, 2, 3 ], [ 1, 4, 5 ], [ 1, 6, 7 ],
       [ 2, 4, 7 ], [ 2, 5, 6 ], [ 3, 4, 6 ], [ 3, 5, 7 ] ],
   isBinary := true, isBlockDesign := true, isSimple := true,
   r := 3, tSubsetStructure := rec( lambdas := [ 1 ], t := 2 ),
   v := 7 ) ]
```

There is, as is very well known, just one such design up to isomorphism. Note
that GAP has printed the value assigned to the variable designs. This output is
a list containing exactly one block design, in DESIGN package format, stored as a
GAP record with properties stored as record components. This output could have
been suppressed by ending the assignment statement with "; ;" instead of ";".

We next classify (but do not display) the 1-(7, 3, 3) designs having no repeated
block and invariant under the group generated by (1, 2)(3, 4).

```
gap> onedesigns:=BlockDesigns(rec( v:=7, blockSizes:=[3],
>      blockMaxMultiplicities:=[1],
>      requiredAutSubgroup:=Group((1,2)(3,4)),
>      tSubsetStructure:=rec(t:=1, lambdas:=[3] ) ) );;
gap> List(onedesigns,AllTDesignLambdas);
[ [ 7, 3 ], [ 7, 3 ], [ 7, 3 ], [ 7, 3 ], [ 7, 3 ],
    [ 7, 3, 1 ] ]
gap> List(onedesigns,d->Size(AutomorphismGroup(d)));
[ 48, 4, 6, 4, 8, 168 ]
```

For a more serious calculation, used in [24], we classify the block designs having 11 points, such that each block has size 4 or 5, and every pair of distinct points is contained in exactly two blocks. This calculation takes about 220 s of CPU-time on a 3.1 GHz PC running Linux.

```
gap> designs:=BlockDesigns(rec(v:=11, blockSizes:=[4,5],
>     tSubsetStructure:=rec(t:=2, lambdas:=[2] ) ) );;
gap> List(designs,BlockSizes);
[ [ 5 ], [ 4, 5 ], [ 4, 5 ], [ 4, 5 ], [ 4, 5 ] ]
gap> List(designs,BlockNumbers);
[ [ 11 ], [ 10, 5 ], [ 10, 5 ], [ 10, 5 ], [ 10, 5 ] ]
gap> List(designs,d->Size(AutomorphismGroup(d)));
[ 660, 6, 8, 12, 120 ]
```

More generally, `BlockDesigns` can construct subdesigns of a given block design, such that the subdesigns each have the same user-specified properties. Here, a *subdesign* of Δ means a block design with the same point set as Δ and whose block multiset is a submultiset of the blocks of Δ. In this case, the default G determining isomorphism is Aut(Δ). For example, given a block design Δ having v points and each of whose blocks has size k, we classify the "parallel classes" of Δ by classifying the subdesigns of Δ that are 1-$(v, k, 1)$ designs (up to the action of Aut(Δ)).

A DESIGN package function closely related to `BlockDesigns` is the function `PartitionsIntoBlockDesigns`, which classifies the partitions of (the block multiset of) a given block design Δ, such that the subdesigns of Δ whose block multisets are the parts of this partition each have the same user-specified properties. For example, given a block design Λ having v points and each of whose blocks has size k, we classify the "resolutions" of Δ by classifying the partitions of Δ into 1-$(v, k, 1)$ subdesigns (up to the action of Aut(Δ)).

3.6.2 The `BlockDesignEfficiency` Function

To test conjectures and rank designs we have classified, we need to be able to compare efficiency measures **exactly**. We can do this using algebraic computation in GAP, as described in [37], and this functionality is included in the most recent release [34] of the DESIGN package.

If `delta` is a 1-(v, k, r) design (in DESIGN package format) with $v > 1$, and `eps` is a positive rational number, then, in DESIGN 1.6, the function call

```
BlockDesignEfficiency(delta,eps)
```

returns a GAP record `eff` (say) having the following components. The component `eff.A` contains the rational number which is the A-efficiency measure of `delta`, `eff.Dpowered` contains the rational number which is the D-efficiency measure of `delta` raised to the power $v - 1$, and `eff.Einterval` is a list $[a, b]$ of non-negative rational numbers such that if E is the E-efficiency measure of `delta` then $a \leq E \leq b$, $b - a \leq$ eps, and if E is rational then $a = E = b$. In addition, the component eff.CEFpolynomial contains the monic polynomial over the rationals whose zeros (counting multiplicities) are the canonical efficiency factors of the design delta.

For example, we calculate the block design efficiency record for one of the 1-$(7, 3, 3)$ designs classified above.

```
gap> eps:=10^(-8);;
gap> delta:=onedesigns[1];
rec( allTDesignLambdas := [ 7, 3 ],
    autGroup := Group([ (1,2), (6,7), (4,5)(6,7),
        (3,4,5)(6,7), (1,6,2,7)(4,5) ]),
    blockNumbers := [ 7 ], blockSizes := [ 3 ],
    blocks := [ [ 1, 2, 3 ], [ 1, 2, 4 ], [ 1, 2, 5 ],
        [ 3, 4, 5 ], [ 3, 6, 7 ], [ 4, 6, 7 ], [ 5, 6, 7 ] ],
    isBinary := true, isBlockDesign := true,
    isSimple := true, r := 3,
    tSubsetStructure := rec( lambdas := [ 3 ], t := 1 ),
    v := 7 )
gap> eff:=BlockDesignEfficiency(delta,eps);;
gap> eff.A;
21/31
gap> eff.Dpowered;
343/2187
gap> eff.Einterval;
[ 1/3, 1/3 ]
gap> Factors(eff.CEFpolynomial);
[ x_1-1, x_1-1, x_1-7/9, x_1-7/9, x_1-7/9, x_1-1/3 ]
```

3.7   Classifying Semi-Latin Squares

We now describe how to classify semi-Latin squares via their duals. Our approach, which is somewhat similar to that of Bailey and Chigbu [7], is implemented in the DESIGN package function SemiLatinSquareDuals, but can be applied with more flexibility using the function BlockDesigns.

The group W_n below will be used to define isomorphism of semi-Latin squares, via their duals. We define

$$W_n := \langle S_n \times S_n, \tau \mid \tau^2 = 1, \tau(a, b)\tau = (b, a) \text{ for all } a, b \in S_n\rangle.$$

Thus W_n is isomorphic to the wreath product $S_n \wr C_2$. If $g \in W_n$ then $g = (a, b)$ or $g = (a, b)\tau$, for some $a, b \in S_n$, and W_n acts on $V := \{1, \ldots, n\}^2$ as follows. For $(i, j) \in V$ and $a, b \in S_n$:

$$(i, j)^{(a,b)} := (i^a, j^b);$$
$$(i, j)^{(a,b)\tau} := (j^b, i^a).$$

Now let S be an $(n \times n)/k$ semi-Latin square, let $S^* = (V, \mathscr{B})$ be the dual of S, and let $g \in W_n$. Define $(S^*)^g := (V, \mathscr{B}^g) = (V, [B^g : B \in \mathscr{B}])$. Then $(S^*)^g$ is the dual of a semi-Latin square T isomorphic to S. Indeed, if $g = (a, b)$ then T can be obtained from S by permuting its rows by a and its columns by b, and if $g = (a, b)\tau$ then T is obtained from S by permuting its rows by a, its columns by b, and then transposing. Conversely, suppose S and T are isomorphic $(n \times n)/k$ semi-Latin squares, with respective duals S^* and T^*. Then T can be obtained from S by applying some row permutation a, some column permutation b, followed possibly by transposing and/or renaming symbols. Then $(S^*)^{(a,b)} = T^*$ if transposing does not take place, and otherwise $(S^*)^{(a,b)\tau} = T^*$. We thus have an action of W_n on the set of duals of $(n \times n)/k$ semi-Latin squares, and the duals of two $(n \times n)/k$ semi-Latin squares X and Y are in the same W_n-orbit if and only if X and Y are isomorphic (as semi-Latin squares); see also [37].

The following GAP function, to be used later, returns a homomorphism from the imprimitive wreath product $S_n \wr C_2$ with block system $\{\{1, \ldots, n\}, \{n + 1, \ldots, 2n\}\}$ onto the group W_n as a permutation group in its action on $V := \{1, \ldots, n\}^2$ as described above. However, the domain of a permutation group in GAP must be a set of positive integers, so the image of the homomorphism is made to be a permutation group on the set $\{1, \ldots, n^2\}$, with i representing the i-th element of V in lexicographic order.

```
gap> L2ActionHomomorphism := function(n)
> local action,tau,W;
> if not IsPosInt(n) then
>     Error("usage: L2ActionHomomorphism( <PosInt> )");
> fi;
> action := function(x,g)
>     # the function which determines the image of
>     #  x  (in {1,...,n^2})  under  g.
>     local i,j,ii,jj;
>     i:=QuoInt(x-1,n)+1;
>     j:=x-(i-1)*n+n;
>     ii:=i^g;
>     jj:=j^g;
>     if ii<=n then
>        return n*(ii-1)+(jj-n);
>     else
>        return n*(jj-1)+(ii-n);
>     fi;
>     end;
> tau:=Product(List([1..n],i->(i,i+n)));
> W:=Group(Concatenation(
>        GeneratorsOfGroup(SymmetricGroup([1..n])),[tau]));
> # so  W = Sn wr C2  in its imprimitive action on  [1..2*n].
> return ActionHomomorphism(
>        W,             # group
>        [1..n^2],      # domain of action
>        action);       # action of group on domain
> end;;
```

We now define a W_n-invariant block design $U_{n,m} = (V, \mathscr{B}_{n,m})$, which contains the dual of every $(n \times n)/k$ semi-Latin square as a subdesign, as long as this dual has no block of multiplicity greater than m. As before, $V := \{1, \ldots, n\}^2$. The block multiset $\mathscr{B}_{n,m}$ consists of all the subsets of V of the form

$$\{(1, 1^g), \ldots, (n, n^g)\},$$

such that $g \in S_n$, and with each such block having multiplicity m. (Note that $U_{n,m}$ is the dual of a certain $(n \times n)/(m(n-1)!)$ semi-Latin square.) The GAP function defined below returns this "universal semi-Latin square dual" $U_{n,m}$ in DESIGN package format.

```
gap> UniversalSemiLatinSquareDual := function(n,m)
> local g,i,block,blocks,U;
> if n<=1 or m<=0 then
>    Error("<n> must be > 1 and <m> must be > 0");
> fi;
> blocks:=[];
> for g in SymmetricGroup([1..n]) do
>    block:=[];
>    for i in [1..n] do
>        Add(block,(i-1)*n+i^g);
>    od;
>    for i in [1..m] do
>        Add(blocks,block);
>    od;
> od;
> U:=BlockDesign(n^2,blocks);
> U.pointNames:=Immutable(Cartesian([1..n],[1..n]));
> return U;
> end;;
```

Observe that a block design $\Delta = (V, \mathscr{B})$ is the dual of an $(n \times n)/k$ semi-Latin square if and only if Δ is a 1-(n^2, n, k) design as well as a subdesign of $U_{n,k}$ (i.e. Δ and $U_{n,k}$ have the same point set and $\mathscr{B}$ is a submultiset of $\mathscr{B}_{n,k}$). We thus obtain the following:

Theorem 3.9 ([37]). *The isomorphism classes of the $(n \times n)/k$ semi-Latin squares are in one-to-one correspondence with the W_n-orbits of 1-(n^2, n, k) subdesigns of $U_{n,k}$. Representatives of these orbits give the duals of isomorphism class representatives of the $(n \times n)/k$ semi-Latin squares.*

This theorem can be adapted to be able to apply the function BlockDesigns in the DESIGN package to construct and classify semi-Latin squares whose duals satisfy certain W_n-invariant properties, and/or whose duals are invariant under a user-specified subgroup of W_n. In particular:

Theorem 3.10 ([37]). *The isomorphism classes of the SOMA(k, n)s are in one-to-one correspondence with the W_n-orbits of the 1-(n^2, n, k) subdesigns of $U_{n,1}$ having the property that any two distinct blocks meet in at most one point.*

Representatives of these orbits give the duals of isomorphism class representatives of the SOMA(k, n)s.

As an application, we find that, up to isomorphism, there are 2799 SOMA(2, 6)s, and just 4 SOMA(3, 6)s (see [31]).

Theorem 3.11 ([37]). *Suppose $n - 1$ divides k and let $\mu := k/(n - 1)$. The isomorphism classes of the uniform $(n \times n)/k$ semi-Latin squares are in one-to-one correspondence with the W_n-orbits of the subdesigns of $U_{n,\mu}$ with the property that any two points having no co-ordinate in common occur together in exactly μ blocks. Representatives of these orbits give the duals of isomorphism class representatives of the uniform $(n \times n)/k$ semi-Latin squares.*

As an application, we find that, up to isomorphism, there are just 10 uniform $(5 \times 5)/8$ semi-Latin squares, and just 277 uniform $(5 \times 5)/12$ semi-Latin squares.

A further, more complicated application is given below. We use the function BlockDesigns to classify, up to isomorphism, the uniform $(6 \times 6)/10$ semi-Latin squares with the property that any two symbols occur together in at most two entries (equivalently, any two blocks in the dual of such a square meet in at most two points). It turns out that there are exactly 98 such semi-Latin squares, and we compute a list L of their duals. We then determine the sizes of the automorphism groups of the elements of L. (The automorphism group of the dual S^* of an $(n \times n)/k$ semi-Latin square S is defined to be the subgroup of W_n that preserves the block multiset of S^*. This is the intersection of W_n with the standard automorphism group of S^* regarded as a block design with no structure on the point set.) The total time taken for this calculation is about 7 min on a 3.1 GHz PC running Linux.

```
gap> n:=6;;
gap> hom:=L2ActionHomomorphism(n);;
gap> W:=Image(hom);;
gap> mu:=2;;
gap> U:=UniversalSemiLatinSquareDual(n,mu);;
gap> rel1:=Set(Orbit(W,[1,2],OnSets));;
gap> rel2:=Difference(Combinations([1..n^2],2),rel1);;
gap> L:=BlockDesigns(rec(v:=n^2,
>       blockDesign:=U,   # we are looking at subdesigns of U
>       blockSizes:=[n],
>       tSubsetStructure:=rec(t:=2, partition:=[rel1,rel2],
>           lambdas:=[0,mu]),
>       blockIntersectionNumbers:=[[[0,1,2]]],
>       isoGroup:=W) );;
gap> A:=List(L,x->Intersection(W,AutomorphismGroup(x)));;
gap> autsizes:=List(A,Size);;
gap> Collected(autsizes);
[ [ 4, 24 ], [ 8, 33 ], [ 12, 4 ], [ 16, 14 ], [ 24, 11 ],
    [ 32, 2 ], [ 48, 2 ], [ 80, 1 ], [ 96, 2 ], [ 144, 1 ],
    [ 192, 2 ], [ 576, 1 ], [ 14400, 1 ] ]
```

A further calculation, taking just 1s, shows that no design in L is resolvable; equivalently, no semi-Latin square whose dual is in L is the superposition of Latin squares.

```
gap> R:=List(L,design->PartitionsIntoBlockDesigns(rec(v:=n^2,
>          blockDesign:=design,
>          blockSizes:=[n],
>          tSubsetStructure:=rec(t:=1, lambdas:=[1]),
>          isoLevel:=0) ) );;
gap> Collected(List(R,Length));
[ [ 0, 98 ] ]
```

3.8 Efficient Semi-Latin Squares as Subsquares of Uniform Semi-Latin Squares

For n a prime power, Bailey [2] gives a construction which produces efficient (although not necessarily optimal [37]) $(n \times n)/k$ semi-Latin squares for all $k > 1$. Efficient (but not known to be optimal) $(6 \times 6)/k$ semi-Latin squares are known for $k = 2, 3, 4, 5, 6$; see [5, 8, 31, 37]. Any uniform $(6 \times 6)/10$ semi-Latin square, such as SLS($PSL_2(5)$), is Schur-optimal, and so is A-, D- and E-optimal. This leaves open the problem of finding efficient $(6 \times 6)/k$ semi-Latin squares, for $k = 7, 8, 9$. To do this, we look at "subsquares" of a uniform $(6 \times 6)/10$ semi-Latin square.

We say that an $n \times n$ semi-Latin square S is a *subsquare* of an $n \times n$ semi-Latin square T if $S = T$ or T is the superposition of S and another $n \times n$ semi-Latin square. Another way of looking at subsquares is via duals. We have S a subsquare of T if and only if the symbol set of S is a subset of the symbol set of T and the dual S^* of S is a 1-(n^2, n, r) subdesign of T^*, for some $r > 0$. In [37], subsquares of uniform semi-Latin squares are investigated, and the following result is proved.

Theorem 3.12 ([37]). *Let $n \geq 3$ and let S be an $(n \times n)/k$ subsquare of a uniform $(n \times n)/t$ semi-Latin square T, such that $t - k < n - 1$. Then*

$$E_S = 1 - t/(k(n-1)) = 1 - \mu(T)/k.$$

Now let $k \in \{7, 8, 9\}$. By Theorem 3.12, each $(6 \times 6)/k$ subsquare of a uniform $(6 \times 6)/10$ semi-Latin square has E-efficiency measure $1 - 2/k$. We obtain an efficient $(6 \times 6)/k$ semi-Latin square Y_k by taking the most A-efficient subsquare of that size of a certain uniform $(6 \times 6)/10$ semi-Latin square Y_{10}. The square Y_{10} was chosen as follows. The list L above of the duals of the 98 uniform $(6 \times 6)/10$ semi-Latin squares with the property that any two distinct symbols occur together in at most two blocks includes the dual of SLS($PSL_2(5)$), whose automorphism group has size 14,400. However, SLS($PSL_2(5)$) has no $(6 \times 6)/9$ subsquare (equivalently it has no Latin square of order 6 as a subsquare). The next largest automorphism group size amongst the 98 dual squares in L is 576, and the one square whose dual in L has an automorphism group of size 576 is chosen as Y_{10}, which does indeed

have $(6 \times 6)/k$ subsquares, for $k = 7, 8, 9$. We have chosen a square whose dual
has a large automorphism group to facilitate the classification of subsquares.

```
gap> f:=First([1..Length(L)],i->Size(A[i])=576);;
gap> Y10star:=L[f];;
gap> autY10star:=A[f];;
gap> StructureDescription(autY10star);
"((A4 x A4) : C2) : C2"
```

We give the square Y_{10} columnwise below:

1	2	3	4	5	6	7	8	9	10
11	12	21	22	31	32	41	42	51	52
13	19	23	25	35	37	47	49	53	59
14	16	27	29	33	40	44	46	57	60
15	18	28	30	34	39	43	45	55	58
17	20	24	26	36	38	48	50	54	56

11	12	13	14	15	16	17	18	19	20
1	2	23	24	33	34	43	44	53	54
3	9	21	27	38	39	45	48	51	57
5	8	22	30	32	36	47	50	55	59
6	7	26	29	31	35	42	49	56	60
4	10	25	28	37	40	41	46	52	58

21	22	23	24	25	26	27	28	29	30
3	4	13	14	35	36	45	46	55	56
1	5	11	17	31	40	43	50	58	60
2	7	15	19	37	39	42	48	52	54
8	10	16	20	33	38	41	47	51	53
6	9	12	18	32	34	44	49	57	59

31	32	33	34	35	36	37	38	39	40
5	6	15	16	25	26	47	48	57	58
4	7	18	20	22	29	41	44	54	55
1	10	12	17	23	28	45	49	51	56
2	9	11	13	24	27	46	50	52	59
3	8	14	19	21	30	42	43	53	60

41	42	43	44	45	46	47	48	49	50
7	8	17	18	27	28	37	38	59	60
6	10	14	15	24	30	32	33	52	56
4	9	11	20	21	26	34	35	53	58
1	3	12	19	22	25	36	40	54	57
2	5	13	16	23	29	31	39	51	55

51	52	53	54	55	56	57	58	59	60
9	10	19	20	29	30	39	40	49	50
2	8	12	16	26	28	34	36	42	46
3	6	13	18	24	25	31	38	41	43
4	5	14	17	21	23	32	37	44	48
1	7	11	15	22	27	33	35	45	47

We now classify the $(6 \times 6)/7$ subsquares of Y_{10} by classifying the 1-$(36, 6, 7)$ subdesigns of Y_{10}^*, up to the action of the automorphism group of Y_{10}^*, as follows.

```
gap> k:=7;;
gap> subdesigns:=BlockDesigns(rec(v:=n^2,
>       blockDesign:=Y10star,
>       blockSizes:=[n],
>       tSubsetStructure:=rec(t:=1, lambdas:=[k]),
>       isoGroup:=autY10star) );;
gap> Length(subdesigns);
150
```

The determination of these 150 subdesigns takes about 7 min on a 3.1 GHz PC running Linux.

We next determine the design(s) in the list subdesigns with the highest A-efficiency measure. There is just one such subdesign, and it also has the highest D-efficiency measure of those in the list. These calculations take about 16 s.

```
gap> eff:=List(subdesigns,BlockDesignEfficiency);;
gap> maxA:=Maximum(List(eff,x->x.A));
18972014997910099125/22524377910796536046
gap> pos:=Filtered([1..Length(eff)],j->eff[j].A=maxA);
[ 65 ]
gap> ForAll([1..Length(eff)],
>       j->eff[pos[1]].Dpowered>=eff[j].Dpowered);
true
```

Let Y_7 be the subsquare of Y_{10} whose dual Y_7^* is the design in the list subdesigns with the highest A-efficiency measure. Then Y_7 is the $(6 \times 6)/7$ semi-Latin square obtained from Y_{10} by removing the $(6 \times 6)/3$ semi-Latin square induced on the symbols

$$\{4, 8, 9, 13, 15, 17, 22, 25, 26, 32, 33, 39, 43, 46, 49, 51, 54, 60\}.$$

The A-efficiency measure of Y_7 is

$$22224360426123258975/25776723339009695896 \approx 0.8622.$$

Similar calculations find a $(6 \times 6)/8$ subsquare of Y_{10} with the highest A-efficiency measure. It also has the highest D-efficiency measure. This subsquare Y_8 can be obtained from Y_{10} by removing the SOMA$(2, 6)$ induced on the symbols

$$\{2, 6, 17, 19, 21, 29, 33, 36, 41, 45, 58, 59\}.$$

The A-efficiency measure of Y_8 is $3643863/4141988 \approx 0.8797$.

Up to isomorphism, we find there is just one $(6 \times 6)/9$ subsquare of Y_{10}. We may take this to be the semi-Latin square Y_9 obtained from Y_{10} by removing the Latin square induced on the symbols

$$\{8, 11, 25, 39, 44, 56\}.$$

The A-efficiency measure of Y_9 is $2968/3323 \approx 0.8932$.

3.9 Some Open Problems

We conclude this chapter with some open problems.

When n is a prime power or $n = 6$, we know precisely the values of μ for which there exists a uniform $(n \times n)/(\mu(n - 1))$ semi-Latin square, but we do not know exactly which values of μ have this property for any other $n > 1$. The first unsettled case is $n = 10$. It is a celebrated computational result that there is no projective plane of order 10 [22, 28], so there do not exist nine MOLS of order 10, and so a uniform $(10 \times 10)/9$ semi-Latin square does not exist. On the other hand, SLS$(PSL_2(9))$ and inflations of this square yield uniform $(10 \times 10)/(9\mu)$ semi-Latin squares for $\mu = 4, 8, 12, 16, \ldots$. Our first question is: do there exist uniform $(10 \times 10)/18$ or $(10 \times 10)/27$ semi-Latin squares?

We have classified certain types of uniform $(6 \times 6)/10$ semi-Latin squares, and found none which is a superposition of Latin squares. Our second question is: does there exist a uniform $(6 \times 6)/10$ semi-Latin square which is a superposition of ten Latin squares?

Finally, are there general constructions for optimal (say E-optimal) $(n \times n)/k$ semi-Latin squares when there do not exist k MOLS of order n and there is no uniform $(n \times n)/k$ semi-Latin square? For example, is every $(n \times n)/(k - 1)$ subsquare of a uniform $(n \times n)/k$ semi-Latin square E-optimal?

Acknowledgements I am grateful to Martin Liebeck for his result of Sect. 3.5.1 and for allowing its inclusion in this chapter. I also thank Eamonn O'Brien for his calculations in MAGMA. The hospitality of the de Brún Centre for Computational Algebra, NUI Galway, during the Fifth de Brún Workshop is gratefully acknowledged.

References

1. R.A. Bailey, Semi-Latin squares. J. Stat. Plan. Inference **18**, 299–312 (1988)
2. R.A. Bailey, Efficient semi-Latin squares. Stat. Sinica **2**, 413–437 (1992)
3. R.A. Bailey, *Association Schemes. Designed Experiments, Algebra and Combinatorics* (Cambridge University Press, Cambridge, 2004)
4. R.A. Bailey, *Design of Comparative Experiments* (Cambridge University Press, Cambridge, 2008)
5. R.A. Bailey, Symmetric factorial designs in blocks. J. Stat. Theor. Pract. **5**, 13–24 (2011)
6. R.A. Bailey, P.J. Cameron, Combinatorics of optimal designs, in *Surveys in Combinatorics 2009*, ed. by S. Huczynska et al. (Cambridge University Press, Cambridge, 2009), pp. 19–73
7. R.A. Bailey, P.E. Chigbu, Enumeration of semi-Latin squares. Discrete Math. **167–168**, 73–84 (1997)
8. R.A. Bailey, G. Royle, Optimal semi-Latin squares with side six and block size two. Proc. Roy. Soc. Lond. Ser. A **453**, 1903–1914 (1997)
9. W. Bosma, J. Cannon, C. Playoust, The Magma algebra system. I. The user language. J. Symbolic Comput. **24**, 235–265 (1997)
10. P.J. Cameron, *Permutation Groups* (Cambridge University Press, Cambridge, 1999)
11. P.J. Cameron, P. Dobcsányi, J.P. Morgan, L.H. Soicher, The External Representation of Block Designs, http://designtheory.org/library/extrep/extrep-1.1-html/
12. P.J. Cameron, L.H. Soicher, Block intersection polynomials. Bull. Lond. Math. Soc. **39**, 559–564 (2007)
13. C.-S. Cheng, R.A. Bailey, Optimality of some two-associate-class partially balanced incomplete-block designs. Ann. Stat. **19**, 1667–1671 (1991)
14. P.E. Chigbu, Semi-Latin Squares: Methods for Enumeration and Comparison, Ph.D. Thesis, University of London, 1996
15. J.H. Conway, R.T. Curtis, S.P. Norton, R.A. Parker, R.A. Wilson, *ATLAS of Finite Groups* (Clarendon Press, Oxford, 1985)
16. P. Dobcsányi, D.A. Preece, L.H. Soicher, On balanced incomplete-block designs with repeated blocks. Eur. J. Comb. **28**, 1955–1970 (2007)
17. R.N. Edmondson, Trojan square and incomplete Trojan square designs for crop research. J. Agric. Sci. **131**, 135–142 (1998)
18. The GAP Group, GAP — Groups, Algorithms, and Programming, Version 4.5.2(beta), 2011, http://www.gap-system.org/
19. A. Giovagnoli, H.P. Wynn, Optimum continuous block designs. Proc. Roy. Soc. Lond. Ser. A **377**, 405–416 (1981)
20. G. James, M. Liebeck, *Representations and Characters of Groups*, 2nd edn. (Cambridge University Press, Cambridge, 2001)
21. P. Kaski, P.R.J. Östergård, *Classification Algorithms for Codes and Designs* (Springer, Berlin, 2006)
22. C.W.H. Lam, L. Thiel, S. Swiercz, The non-existence of finite projective planes of order 10. Can. J. Math. **41**, 1117–1123 (1989)
23. J.H. van Lint, R.M. Wilson, *A Course in Combinatorics* (Cambridge University Press, Cambridge, 1992)
24. J.P. McSorley, L.H. Soicher, Constructing *t*-designs from *t*-wise balanced designs. Eur. J. Comb. **28**, 567–571 (2007)
25. N.C.K. Phillips, W.D. Wallis, All solutions to a tournament problem. Congr. Numerantium **114**, 193–196 (1996)
26. D.A. Preece, G.H. Freeman, Semi-Latin squares and related designs. J. Roy. Stat. Soc. Ser. B **45**, 267–277 (1983)
27. D.A. Preece, N.C.K. Phillips, Euler at the bowling green. Utilitas Math. **61**, 129–165 (2002)
28. D.J. Roy, Confirmation of the Non-existence of a Projective Plane of Order 10, M.Sc. Thesis, Carleton University, Ottawa, 2011

29. K.R. Shah, B.K. Sinha, *Theory of Optimal Designs*. Lecture Notes in Statistics, vol. 54 (Springer, Berlin, 1989)
30. L.H. Soicher, On the structure and classification of SOMAs: generalizations of mutually orthogonal Latin squares. Electron. J. Comb. **6**, R32 (1999), Printed version: J. Comb. **6**, 427–441 (1999)
31. L.H. Soicher, SOMA Update, http://www.maths.qmul.ac.uk/~leonard/soma/
32. L.H. Soicher, Computational group theory problems arising from computational design theory. *Oberwolfach Rep.* **3** (2006), Report 30/2006: Computational group theory, 1809–1811
33. L.H. Soicher, More on block intersection polynomials and new applications to graphs and block designs. J. Comb. Theor. Ser. A **117**, 799–809 (2010)
34. L.H. Soicher, The DESIGN package for GAP (2011), Version 1.6, http://designtheory.org/software/gap_design/
35. L.H. Soicher, The GRAPE package for GAP, Version 4.5, 2011, http://www.maths.qmul.ac.uk/~leonard/grape/
36. L.H. Soicher, Uniform semi-Latin squares and their Schur-optimality. J. Comb. Des. **20**, 265–277 (2012)
37. L.H. Soicher, Optimal and efficient semi-Latin squares. J. Stat. Plan. Inference **143**, 573–582 (2013)
38. C.-Y. Suen, I.M. Chakravarti, Efficient two-factor balanced designs. J. Roy. Stat. Soc. Ser. B **48**, 107–114 (1986)
39. M.N. Vartak, The non-existence of certain PBIB designs. Ann. Math. Stat. **30**, 1051–1062 (1959)

LECTURE NOTES IN MATHEMATICS

Edited by J.-M. Morel, B. Teissier; P.K. Maini

Editorial Policy (for the publication of monographs)

1. Lecture Notes aim to report new developments in all areas of mathematics and their applications - quickly, informally and at a high level. Mathematical texts analysing new developments in modelling and numerical simulation are welcome.
   Monograph manuscripts should be reasonably self-contained and rounded off. Thus they may, and often will, present not only results of the author but also related work by other people. They may be based on specialised lecture courses. Furthermore, the manuscripts should provide sufficient motivation, examples and applications. This clearly distinguishes Lecture Notes from journal articles or technical reports which normally are very concise. Articles intended for a journal but too long to be accepted by most journals, usually do not have this "lecture notes" character. For similar reasons it is unusual for doctoral theses to be accepted for the Lecture Notes series, though habilitation theses may be appropriate.

2. Manuscripts should be submitted either online at www.editorialmanager.com/lnm to Springer's mathematics editorial in Heidelberg, or to one of the series editors. In general, manuscripts will be sent out to 2 external referees for evaluation. If a decision cannot yet be reached on the basis of the first 2 reports, further referees may be contacted: The author will be informed of this. A final decision to publish can be made only on the basis of the complete manuscript, however a refereeing process leading to a preliminary decision can be based on a pre-final or incomplete manuscript. The strict minimum amount of material that will be considered should include a detailed outline describing the planned contents of each chapter, a bibliography and several sample chapters.
   Authors should be aware that incomplete or insufficiently close to final manuscripts almost always result in longer refereeing times and nevertheless unclear referees' recommendations, making further refereeing of a final draft necessary.
   Authors should also be aware that parallel submission of their manuscript to another publisher while under consideration for LNM will in general lead to immediate rejection.

3. Manuscripts should in general be submitted in English. Final manuscripts should contain at least 100 pages of mathematical text and should always include

   – a table of contents;
   – an informative introduction, with adequate motivation and perhaps some historical remarks: it should be accessible to a reader not intimately familiar with the topic treated;
   – a subject index: as a rule this is genuinely helpful for the reader.

   For evaluation purposes, manuscripts may be submitted in print or electronic form (print form is still preferred by most referees), in the latter case preferably as pdf- or zipped psfiles. Lecture Notes volumes are, as a rule, printed digitally from the authors' files. To ensure best results, authors are asked to use the LaTeX2e style files available from Springer's web-server at:

   ftp://ftp.springer.de/pub/tex/latex/svmonot1/ (for monographs) and
   ftp://ftp.springer.de/pub/tex/latex/svmultt1/ (for summer schools/tutorials).

Additional technical instructions, if necessary, are available on request from lnm@springer.com.

4. Careful preparation of the manuscripts will help keep production time short besides ensuring satisfactory appearance of the finished book in print and online. After acceptance of the manuscript authors will be asked to prepare the final LaTeX source files and also the corresponding dvi-, pdf- or zipped ps-file. The LaTeX source files are essential for producing the full-text online version of the book (see http://www.springerlink.com/openurl.asp?genre=journal&issn=0075-8434 for the existing online volumes of LNM). The actual production of a Lecture Notes volume takes approximately 12 weeks.

5. Authors receive a total of 50 free copies of their volume, but no royalties. They are entitled to a discount of 33.3 % on the price of Springer books purchased for their personal use, if ordering directly from Springer.

6. Commitment to publish is made by letter of intent rather than by signing a formal contract. Springer-Verlag secures the copyright for each volume. Authors are free to reuse material contained in their LNM volumes in later publications: a brief written (or e-mail) request for formal permission is sufficient.

Addresses:
Professor J.-M. Morel, CMLA,
École Normale Supérieure de Cachan,
61 Avenue du Président Wilson, 94235 Cachan Cedex, France
E-mail: morel@cmla.ens-cachan.fr

Professor B. Teissier, Institut Mathématique de Jussieu,
UMR 7586 du CNRS, Équipe "Géométrie et Dynamique",
175 rue du Chevaleret
75013 Paris, France
E-mail: teissier@math.jussieu.fr

For the "Mathematical Biosciences Subseries" of LNM:

Professor P. K. Maini, Center for Mathematical Biology,
Mathematical Institute, 24-29 St Giles,
Oxford OX1 3LP, UK
E-mail : maini@maths.ox.ac.uk

Springer, Mathematics Editorial, Tiergartenstr. 17,
69121 Heidelberg, Germany,
Tel.: +49 (6221) 4876-8259

Fax: +49 (6221) 4876-8259
E-mail: lnm@springer.com